3/91

WITHDRAWN

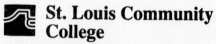

St. Louis Community College

Forest Park
Florissant Valley
Meramec

Instructional Resources
St. Louis, Missouri

St. Louis Community College
at Meramec
Library

The Asian elephant: ecology and management

Cambridge Studies in Applied Ecology and Resource Management

The rationale underlying much recent ecological research has been the necessity to understand the dynamics of species and ecosystems in order to predict and minimise the possible consequences of human activities. As the social and economic pressures for development rise, such studies become increasingly relevant, and ecological considerations have come to play a more important role in the management of natural resources. The objective of this series is to demonstrate how ecological research should be applied in the formation of rational management programmes for natural resources, particularly where social, economic or conservation issues are involved. The subject matter will range from single species where conservation or commercial considerations are important to whole ecosystems where massive perturbations like hydro-electric schemes or changes in land-use are proposed. The prime criterion for inclusion will be the relevance of the ecological research to elucidate specific, clearly defined management problems, particularly where development programmes generate problems of incompatibility between conservation and commercial interests.

Editorial Board

Dr G. Caughley. Division of Wildlife and Rangelands Research, CSIRO, Australia

Dr S. K. Eltringham. Department of Applied Biology, University of Cambridge, UK

Dr J. Harwood. Sea Mammal Research Unit, Natural Environment Research Council, Cambridge, UK

Dr D. Pimentel. Department of Entomology, Cornell University, USA

Dr A. R. E. Sinclair. Institute of Animal Resource Ecology, University of British Columbia, Canada

Dr M. P. Sissenwine. Northeast Fishery Center, National Marine Fisheries Service, Woods Hole, USA

Also in the series

Graeme Caughley, Neil Shephard & Jeff Short (eds.)   *Kangaroos: their ecology and management in the sheep rangelands of Australia*

P. Howell, M. Lock & S. Cobb (eds.)   *The Jonglei canal: impact and opportunity*

Robert J. Hudson, K. R. Drew & L. M. Baskin (eds.)   *Wildlife production systems: economic utilization of wild ungulates*

M. S. Boyce   *The Jackson elk herd: intensive wildlife management in North America*

Mark R. Stanley Price   *Animal re-introductions: the Arabian oryx in Oman*

THE ASIAN ELEPHANT: ECOLOGY AND MANAGEMENT

R. Sukumar
*Centre for Ecological Sciences,
Indian Institute of Science, Bangalore, India*

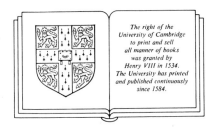

CAMBRIDGE UNIVERSITY PRESS
Cambridge
New York   Port Chester
Melbourne   Sydney

Published by the Press Syndicate of the University of Cambridge
The Pitt Building, Trumpington Street, Cambridge CB2 1RP
40 West 20th Street, New York, NY 10011, USA
10 Stamford Road, Oakleigh, Melbourne 3166, Australia

© Cambridge University Press 1989

First published 1989

Printed in Great Britain at
the University Press, Cambridge

British Library cataloguing in publication data
Sukumar, R.
The Asian elephant: ecology and management.
1. Indian elephants. Ecology
I. Title
599.6'1045

Library of Congress cataloguing in publication data
Sukumar, R.
Ecology and management of the Asian elephant / R. Sukumar
    p.   cm. – (Cambridge studies in applied ecology and resource management)
Bibliography: p.
ISBN 0-521-36080-3
1. Asian elephant – Ecology. 2. Wildlife management – Asia.
I. Title. II. Series.
QL737.P98S95   1989
599.6'1'095—dc20

ISBN 0 521 36080 3

To the memory of my grandfather
R. Kantham

CONTENTS

Foreword by Dr M. S. Swaminathan	*page*	xiii
Acknowledgements		xv
1	**The historical background**	**1**
1.1	The period before Christ	2
1.2	The period after Christ	4
1.3	Elephants captured during the nineteenth and twentieth centuries	8
2	**Status and distribution of the Asian elephant**	**10**
2.1	Indian sub-continent	11
2.1.1	*Southern India*	11
2.1.2	*Central India*	15
2.1.3	*Northwestern India and Nepal*	16
2.1.4	*Northeastern India and Bhutan*	17
2.1.5	*Bangladesh*	18
2.2	Continental Southeast Asia	18
2.2.1	*China*	18
2.2.2	*Burma*	18
2.2.3	*Thailand*	20
2.2.4	*Laos, Kampuchea and Vietnam*	22
2.2.5	*Malaysia*	22
2.3	Island Asia	24
2.3.1	*Sri Lanka*	24
2.3.2	*Andaman islands (India)*	26
2.3.3	*Sumatra (Indonesia)*	26
2.3.4	*Borneo (Malaysia and Indonesia)*	28

2.4	Summary of population estimates	29
2.5	Conservation issues	32
2.5.1	*Exploitation of habitat*	32
2.5.2	*Shifting cultivation*	32
2.5.3	*Spread of agriculture*	33
2.5.4	*Hydro-electric and irrigation dams*	34
2.5.5	*Capture of elephants*	35
2.5.6	*Hunting of elephants*	35
2.5.7	*Crop raiding and manslaughter by elephants*	36
2.5.8	*Political unrest and war*	37
2.5.9	*Legal and administrative inadequacies*	37
3	**The main study area and study methods**	**39**
3.1	The study area	40
3.1.1	*Hydrology*	40
3.1.2	*Climate*	42
3.1.3	*Land-use pattern*	43
3.1.4	*Vegetation types*	44
3.1.5	*Mammals*	50
3.1.6	*The people*	53
3.2	Study methods	54
3.2.1	*Registration of elephants*	54
3.2.2	*Density estimates in different zones*	54
3.2.3	*Ageing techniques*	54
3.2.4	*Observations on feeding*	55
3.2.5	*Primary production of grass*	56
3.2.6	*Herbivore biomass and grass consumption*	56
3.2.7	*Damage to trees*	57
3.2.8	*Damage to crops*	57
3.2.9	*Chemical analysis*	58
3.2.10	*Collection of records*	58
4	**Movement and habitat utilization**	**60**
4.1	Elephant densities and seasonal distribution patterns	60
4.2	Movement pattern of different clans	62
4.3	Home range sizes and factors influencing them	63
4.4	Inter-annual differences in movement pattern	65
4.5	A long-term perspective on elephant movement patterns	65
4.6	Seasonal use of habitat types	66
4.7	Movement in relation to foraging and availability of water	67

5	**Feeding and nutrition**	**69**
5.1	The plants and parts eaten	69
5.2	Proportions of browse and grass in the diet	71
5.3	Proportions of different browse plants in the diet	74
5.4	Quantity of forage consumed	78
5.5	Drinking	79
5.6	Nutrition and foraging	80
5.6.1	*Anatomy and physiology*	81
5.6.2	*Palatability and nutritive value of foods*	82
5.6.3	*Plant secondary compounds*	84
5.6.4	*Browse and grass in the optimal diet*	85
6	**Impact on the vegetation and carrying capacity**	**86**
6.1	Primary production of grasses	87
6.2	Biomass of mammalian herbivores and grass consumption	89
6.3	Impact on woody vegetation	91
6.4	Inter-specific competition, resource limitation and carrying capacity	97
6.5	Elephant–vegetation interaction and ecosystem dynamics	99
7	**Crop raiding by elephants**	**108**
7.1	The crops cultivated	108
7.2	The crops consumed and patterns of feeding	111
7.3	Frequency and seasonality of raiding	114
7.4	Group sizes of raiding elephants	116
7.4.1	*Mean monthly raiding sizes of bulls and herds in villages*	116
7.4.2	*Frequency of unit group sizes*	119
7.5	Quantity consumed from crop fields	120
7.5.1	*Rate of foraging in crop fields*	120
7.5.2	*Quantity and proportion of crops in the diet*	120
7.6	Quantity of crops damaged and the economic loss	123
7.7	Methods to deter elephants, and their behavioural responses	125
7.8	Crop raiding in rain forest habitats	126
7.9	Causes of crop raiding	130
7.9.1	*Raiding in relation to movement patterns*	131
7.9.2	*Competition for water*	131
7.9.3	*Reduction and fragmentation of natural habitat*	132
7.9.4	*Degradation of habitat*	133
7.9.5	*Palatability and nutritive value of crops*	133

8	**Manslaughter by elephants**	**135**
8.1	The people killed	135
8.2	The elephants responsible	136
8.3	Circumstances of encounter	137
8.4	Causes of aggression	140
9	**Habitat manipulation by people**	**143**
9.1	Reduction in habitat area and fragmentation	143
9.2	Shifting cultivation	144
9.3	Extraction of plant products	145
9.3.1	*Timber felling*	145
9.3.2	*Bamboo extraction*	147
9.3.3	*Grass extraction*	149
9.3.4	*Minor forest produce collection*	149
9.4	Grazing by domestic livestock	150
9.5	Fire	151
9.6	Forest plantations	154
9.7	Food availability in primary versus secondary rain forest	157
9.8	Carrying capacity of primary versus secondary vegetation	157
9.9	Creation of water reservoirs	159
9.9.1	*The Parambikulam – Aliyar Project in southern India*	160
9.9.2	*The Mahaweli Ganga Development Project in Sri Lanka*	160
10	**Elephant slaughter by people**	**165**
10.1	Elephant populations susceptible to poaching	165
10.2	Causes of death in elephants	166
10.3	Number, age frequency and mean tusk weight of poached elephants	168
10.4	The people responsible	169
10.5	Poaching and the ivory trade in India	171
11	**Population dynamics**	**174**
11.1	Age and sex structure	175
11.2	Fertility	177
11.2.1	*Age at first and last calving*	177
11.2.2	*Birth rate and the mean calving interval*	179
11.3	Mortality	182
11.3.1	*Numbers found dead in the field*	183
11.3.2	*Life-table from ages at death*	185

11.3.3	*Instantaneous mortality rate from the standing age distribution*	186
11.4	Population modelling	188
11.4.1	*Trends in population growth*	189
11.4.2	*Trends in age distribution*	191
11.4.3	*Trends in the adult sex ratio*	193
11.4.4	*Possible effect of a disparate adult sex ratio on fertility*	195
11.4.5	*A test of the validity of the model: predicted and observed trends in the proportion of adult males*	197
11.5	Demographic condition of the population	197
12	**Conservation and management**	**202**
12.1	Conservation of the elephant	202
12.1.1	*Minimum viable population size*	202
12.1.2	*Minimum viable area and habitat integrity*	205
12.1.3	*Maintenance of habitat quality*	208
12.1.4	*Reduction of poaching*	210
12.1.5	*Captive breeding programmes*	210
12.2	Protection of human interests	211
12.2.1	*Elephant-proof barriers*	211
12.2.2	*Discouraging and chasing elephants*	214
12.2.3	*Culling, capture and translocation of elephants*	215
12.2.4	*Agricultural planning*	218
12.2.5	*Social security schemes*	218
	Appendices	
I	Estimation of seasonal elephant densities by ground transects	219
II	Growth relationships and field methods of ageing elephants	224
III	Nutritive value of food plants	227
	References	230
	Index	241

FOREWORD

Elephants have fascinated humankind for millennia. On the Indian sub-continent the two species have enjoyed a particularly close relationship. Elephants figure prominently among the stone age rock paintings of Narmada Valley, and on the seals of the Mohanjodaro civilization. It was in India that elephants were tamed, and helped in the agricultural colonization of the wet valleys of the Ganga and Brahmaputra. They were prized as beasts of war and Indian kings maintained them in thousands for their armies. But man's crop fields provide a most attractive source of food for the elephant and the two must have come into intense conflict since the beginning of agriculture. That is why the ancient Indian manual of statecraft, the Arthasastra, prescribes both the eradication of elephants from river valleys and their strict protection in forests on the borders of the kingdom.

The interaction of elephant and man in the mosaic of natural and man-modified habitats that cover the country is therefore a theme of particular fascination. Dr Sukumar addresses it brilliantly in this study of one of the largest surviving populations of the Indian elephant. He traces how crop raiding fits in with the overall habitat utilization pattern of this population, and carefully documents the ensuing elephant–human conflict. He goes on to make very practical suggestions as to how this conflict may be ameliorated, if not fully resolved. Such prescriptions are of great value in our quest to reconcile imperatives of development with conservation. I therefore have great pleasure in introducing this study, and hope that others in the same vein will follow to help us evolve meaningful strategies for conservation of biological diversity in the developing world. Such studies are essential to give meaning and content to the concept of sustainable development. We owe Dr Sukumar a deep debt of gratitude for this labour of love for the human–elephant partnership.

M. S. SWAMINATHAN
(*President, IUCN*)

ACKNOWLEDGEMENTS

So many people and institutions have helped me during the past nine years that I fear, as I begin to acknowledge their services, my memory might fail in mentioning each one of them. I wish to express my deep gratitude to everyone who has been involved in the completion of this work.

Madhav Gadgil stimulated my interest in elephants, taught, guided and encouraged me during my tenure as his student. My parents bought me a vehicle for the field work and accepted the numerous financial liabilities I imposed upon them.

World Wide Fund for Nature-International (WWF) provided financial support for much of the field work through the Asian Elephant Specialist Group (AESG) of the International Union for Conservation of Nature and Natural Resources (IUCN). M. A. Partha Sarathy (WWF-India and Chairman, IUCN Commission on Education) provided the much needed thrust and local support. J. C. Daniel (Curator, Bombay Natural History Society and former Chairman of the AESG) also strongly supported my work. The Salim Ali Conservation Fund of the Bombay Natural History Society provided a grant for the preliminary field survey.

The Forest Departments of Tamilnadu and Karnataka were extremely helpful to me and supportive of logistics in the field. I must mention, in particular, T. Achaya (Chief Conservator of Forests), K. Venkatakrishnan, K. Shanmuganathan, P. Padmanabhan and M. K. Appayya (Chief Wildlife Wardens), C. R. Thirumurthy, M. Ramachandran, S. Sarangapani, P. K. Devaiah, R. K. Torvi and P. Srinivas (Divisional Forest Officers), the late Chidambaram, Premnath and M. K. Baig (Forest Range Officers). V. Krishnamurthy (Forest Veterinary Officer) freely made available to me all his records and expertise on elephants. C. Jeganthan and B. S. Gopala Rao (Veterinary Surgeons) also provided valuable help.

Acknowledgements

The chemical analysis of plants was made possible with assistance from Kantha Shurpalekar, Octobel Sundaravalli, Arun Chandrasekhar and Shenoy (Central Food Technological Research Institute, Mysore), and C. Vinutha and H. S. Satyanarayana (Indian Institute of Science). The carbon-isotope analysis of bone collagen was undertaken in collaboration with S. K. Bhattacharya, R. V. Krishnamurthy and R. Ramesh at the Physical Research Laboratory, Ahmedabad. V. J. Nair, N. C. Rathakrishnan, A. N. Henry, Srikumar, C. N. Mohanan and Bhargavan of the Botanical Survey of India, Coimbatore, identified many of the plant specimens I collected.

R. Selvakumar accompanied me on the preliminary surveys, hosted me at Mudumalai and shared his observations with me. A. N. Jagannatha Rao equipped me and my vehicle with his exclusive expertise for jungle survival and this prevented what could have been frustrating moments during field work. The late Siddhartha Buch introduced me to wildlife photography and kept his darkroom (and also his heart) always open. M. Krishnan freely shared with me his experiences with elephants. The support of S. T. Baskaran and Tilaka, and V. Srikumar and Shyamala at Coimbatore was welcome when I had to temporarily migrate from the jungles into a nearby city.

The staff of the Centre of Ecological Sciences, the Centre for Theoretical Studies and the Microbiology and Cell Biology Laboratory, Indian Institute of Science, provided a stimulating academic atmosphere and vital logistic support. I benefited greatly from discussions with my colleagues N. V. Joshi, P. V. Nair, S. Narendra Prasad, Raghavendra Gadagkar, Sulochana Gadgil and Fr. C. J. Saldanha. I must thank, in particular, N. V. Joshi for his help in data analysis on the computer. T. Srikantha, Banumathi Ramachandran, N. Ramesh, S. Sankar and V. V. Krishna responded generously to the numerous demands I made on their time.

I have been able to provide a broad perspective on elephants in Asia only because of the work of members of the Asian Elephant Specialist Group of IUCN. I must thank, in particular, D. K. Lahiri Choudhury, Lyn de Alwis, Ullas Karanth, Peter Jackson, Charles Santiapillai, Jacob Cheeran and the late S. P. Shahi for many discussions and unpublished information they passed on to me. The Action Plan for Asian Elephant Conservation by C. Santiapillai was invaluable in updating the chapter on elephant status and distribution. IUCN and WWF-International permitted me to quote freely from their unpublished project reports.

While I was preparing the manuscript, Aviva Patel, Jayakumar Radhakrishnan, H. S. Suresh, R. Arumugam and Nitin Rai helped me to update the information contained in my doctoral thesis. S. Nirmala composed the

Acknowledgements

typescript, A. V. Narayan and M. J. Venugopal prepared the figures and A. G. Ryan the photographs for publication.

Setty spotted elephants, heard elephants, smelt elephants and tracked elephants for three years. His presence made field work easier.

My wife Sudha has patiently endured the marathon thesis and book writing. Our little daughter Gitanjali allowed me to complete the book in spite of interrupting me every five minutes and insisting on sitting on my lap while I wrote.

If I have failed to acknowledge anyone, it only shows that the memory of an elephant researcher is not as good as that of an elephant.

R. Sukumar
Bangalore, India

1

The historical background

An object of worship, a target of hunters, a beast of burden, a burden to the people, gentle in captivity, dangerous in the wild, the pride of kings, the companion of mahouts, a machine of war, an envoy of peace, loved, feared and hated, the elephant has had a glorious and an infamous association with man in Asia. For its sheer contrast and splendour, this association is unequalled by any other interaction between animal and man in the world.

The Asian elephant (*Elephas maximus*) once held sway over a vast region from the Tigris–Euphrates in West Asia eastward through Persia into the Indian sub-continent, South and Southeast Asia including the islands of Sri Lanka, Java, Sumatra and Borneo, and into China northwards up to the Yangtze-Kiang. It has disappeared entirely from West Asia, Persia, Java and most of China. Its distribution over the Indian sub-continent and Southeast Asia is restricted largely to forested hilly tracts, which are usually the last habitats to be taken over by human settlements.

The current status of the Asian elephant, and its conservation problems, can best be understood in the context of its historical association with people. I therefore begin with an overview of the elephant's interaction with people through the centuries, resulting in its depletion in the wild. The historical account is confined to the Indian sub-continent, for which the literature is more extensive and better known than for any other region. This is followed by a review of the elephant's present-day distribution and conservation problems. In subsequent chapters I discuss the ecology of the elephant in the wild and elephant–human interaction. Finally, suggestions for conserving the elephant while minimizing its conflict with people are made.

1.1 The period before Christ

The long and diverse evolutionary history of the Proboscidea, beginning with *Moeritherium* from the Eocene, has left only two living species, the African elephant (*Loxodonta africana*) and the Asian elephant (*Elephas maximus*). Man has been a predator of the Proboscidea since the Pleistocene. Pre-historic hunter–gatherers presumably viewed the elephant only as a source of meat, bones, hide and ivory. Stone Age hunters certainly killed elephants and their contemporaries, the mammoths and the mastodons, as revealed by archaeological remains and cave paintings in Europe, Africa, and Asia (references in Carrington 1958 and Freeman 1980). The earliest indications of domestication of the elephant are the engravings on seals of the Indus valley civilization, dated as third millenium BC (Rao 1957). There were probably wild elephants in this area during that period. The Dravidian people of the Indus civilization were displaced into southern India by the so-called Aryan invaders during the third millenium BC. The settlement of the Indo-Gangetic plains must have accelerated the reduction of natural vegetation and of the elephant population. Ancient literature such as the Rig Veda (twentieth to fifteenth century BC), the Upanishads (ninth to sixth century BC) and Gajasastra (Sanskrit for 'elephant lore') record details of the elephant's distribution, its life and habits (mixed with mythology, exaggeration and imagination) and instructions on its capture, training and maintenance. The Gajasastra, attributed to Palakapya (sixth to fifth century BC), mentions the presence of elephants in practically the whole of India, including the present states of Rajasthan, Punjab, Gujarat, Madhya Pradesh, Andhra and other places from which they have now disappeared. The conflict between elephants and people for cultivated crops, which must have begun with the times of shifting cultivation, would have intensified when settled agriculture became established in river valleys. In an obvious reference to crop raiding, the Gajasastra records that elephants devastated the kingdom of Anga ruled by Romapada.

We can only speculate that the capture and training of elephants, which was practised by the Indus people, became an occupation of the Aryan people. To this day one can see the traditional difference in the method of capturing elephants. In the north the Aryans captured entire elephant herds in stockades, whereas in the south the Dravidians captured them singly or in small numbers in pits (Stracey 1963). Once the elephant was domesticated and put to the service of man, it triggered the decline of its wild relatives. The clearing of the wilderness was made easier by using elephants to fell and transport timber. Trained elephants became indispensable in capturing and domesticating wild elephants.

The next logical step must have been to use this large beast as a war machine. It is not known when the elephant was first used in war. In the Mahabharatha (c. 1100 BC) there is mention of an elephant named Aswathamma, which was killed in a battle between the Pandavas and the Kauravas. Cyrus the Great of Persia was killed in a battle (530 BC) with the Derbikes, who were supplied with Indian elephants. One of the first documented instances in which an army with elephants suffered a decisive defeat was the famous battle (326 BC), between Alexander the Great and the elephant army of Porus, on the banks of the Jhelum. This does not seem to have decreased the passion of kings for possessing war elephants; elephant armies continued to exist in the Indian sub-continent for another twenty centuries.

Chandragupta, the Mauryan emperor who defeated Alexander's successor Sellucus Nikator, had an army of 9000 elephants; other rulers in the sub-continent had between them at least 5000 elephants during this time. The Mauryan kingdom imported elephants from other regions of the sub-continent and probably also from Sri Lanka (see the testimony of Megasthenes referred to by Digby (1971); however, Trautmann (1982) says that elephants from Ceylon, i.e. Sri Lanka, did not go to the Mauryans but to Kalinga or Orissa). Thus in northern India a trade in elephants was set up from at least the third century BC.

During the second and first millennia BC, the kings and chieftains of the more advanced Aryan civilization, cultivating the river valleys, might have introduced a taboo on killing elephants for meat; elephants were more useful alive in their armies. Supplies of elephants may have come largely from hill forests inhabited by shifting cultivators and hunter–gatherers, who may have gradually ceased consuming elephant meat and, instead, captured elephants for the Aryan armies. Kautilya's Arthasastra (c. 300 BC to AD 300), a manual of statecraft, advised that elephants were to be eliminated from river valleys under settlement but preserved in the outer hill forests. It recommended setting up elephant sanctuaries, on the periphery of the kingdom, which were to be patrolled by guards. Anyone killing an elephant within the sanctuary was to be put to death. Interestingly, the Arthasastra also instructed that elephant calves, elephants with small tusks, tuskless males, diseased elephants, female elephants with young and suckling females were not to be captured, but a 20-year old elephant should be captured (see translations by Shamasastry 1960 and Kengle 1972). This effectively meant that only adult male tuskers could be captured; in a sense this was also a prudent way of harvesting the elephant population (discussed in Chapters 11 and 12). No doubt, these instructions were not always practised; we know that entire herds were captured by the *kheddah* method of driving them into stockades.

1.2 The period after Christ

Information on certain aspects, such as elephant distribution or numbers of domestic elephants, is sketchy until the tenth century. The Tamil Sangam literature of southern India has many interesting references to the elephant's habits (Varadarajaiyer 1945). It mentions the presence of elephants in the Tirupati hills of Andhra, where they are now absent. Many tribal chieftains owned considerable numbers of elephants. A pointed reference is made to the abundance of male elephants (no doubt, tuskers) in the hilly country (the Western Ghats of Tamilnadu and Kerala) ruled by the Cheras.

The worship of the elephant god, Ganesha, which originated in the third or fourth century, must have created a strong ethos against the killing of elephants. Ganesha worship reached the Tamil land only during the Pallava regime of the seventh century. Hill tribes in the south certainly hunted elephants for their tusks and consumed elephant meat, probably before this time.

The role of the ivory trade in the decline of the Asian elephant cannot be clearly evaluated. Ivory objects are known from pre-dynastic Egypt, Assyria, Greece and Rome, and later mediaeval Europe and the Islamic countries (Freeman 1980). The principal centres of ivory carving in the East were China, Japan and India. The supply of local Indian ivory was never sufficient to meet the demands of the trade. The harvest of Indian elephants for ivory must have been relatively low because only male elephants carried tusks and tuskers were valued for armies. Thus, African ivory from Ethiopia was imported by India from the 6th century BC onwards (Warmington 1974). Part of this ivory could have been simply in transit to other regions, particularly China. Ivory and ivory objects were also supplied by India to Greece and Rome. The rise of the Babylonian and Persian civilizations across the land-routes between India and the west created a more extensive trade in Indian than in African ivory (Warmington 1974). The Roman trade in ivory was particularly large. How much of the Indian ivory supplied to the west was carved African ivory that had been imported is not known. Ivory from Indian elephants was certainly exported; the best quality ivory is reputed to have come from elephants in Orissa.

Whatever the impact of the ivory trade on elephant populations, the spread of settled agriculture and captures must have eliminated them from large tracts of the Indo-Gangetic plains and river basins in peninsular India by the tenth century. From the eleventh century onwards the rulers in the north obtained few of their war elephants from wild stocks in their own region.

The possession of a large *pil-khana* (elephant stable) was a matter of prestige to the northern rulers. An elephant army was still a psychological force to reckon with in direct battle. The Sultans of Delhi, who used elephants, lost their only battle to a foreign and elephantless army in AD 1398, after a successful reign extending over two centuries. The Ghaznavid kingdom had 1670 war elephants in AD 1031 (Digby 1971). This dropped considerably during the period of the Delhi Sultanate (AD 1192–1398). At the height of their power (*c.* AD 1340) the Delhi Sultanate possessed 3000 elephants, of which only 750–1000 were war elephants, the rest being young elephants or those unfit for battle. Most of the elephants in the Sultanate seemed to have come as captures from enemies in southern India, as tribute from subordinate rulers or imports from various regions including eastern Bengal, Sri Lanka and Pegu in lower Burma (Digby 1971). For instance, Sultan Malik Kafur received as tribute or captured a total of 512 elephants during his third Deccan expedition. The trade in elephants was a complex affair; for instance, elephants from Sri Lanka were exported through southern India and Gujarat, whereas those from Pegu may have been sent through Bengal and Sri Lanka. The import of elephants by the Sultanate was not necessarily due to a total absence of wild stocks from northern India, although it does imply serious depletion. Elephants from certain regions, such as Sri Lanka, were imported because they were considered especially suited for use in war. When the Sultanate was overthrown in AD 1398 by Amir Timur, the *pil-khana* had a mere 120 war elephants.

The invention of gunpowder reduced the effectiveness of the elephant in war, although elephants certainly fought in direct battle lines along with explosive weapons during the fourteenth century. Once rapidly firing artillery and hand guns came into use during the sixteenth century, the elephant ceased to be of any value in direct battle. It continued to be used as a perch for the commander, a siege engine and a haulier of war materials until the nineteenth century.

References to the distribution of wild elephants again surface from the writings of the Moghul rulers of the sixteenth and seventeenth centuries (Salim Ali 1927). Some of these elephants may have been in isolated populations, but others were certainly part of a more widespread distribution over central India adjoining the present elephant range in Orissa. Emperor Babur (AD 1526–30) noted that the elephant 'inhabits the district of Kalpi and the higher you advance thence towards the east, the more do the elephants increase in number'. Abul Fazl's Ain-i-Akbari, a chronicle of the times of Emperor Akbar (AD 1556–1605), records elephants in many parts of central India including Marwar, Chanderi, Satwas, Bijagarh, Raisen and Panna. Jehangir (AD 1605–27) described an elephant hunt in Dohad in the

Panchmahal hills. The elephant hunts of the Moghul rulers seem to refer only to capture and not to killing of elephants, although other animals were killed during their sport hunts. There was hardly any breeding of elephants in captivity.

The Moghuls continued to build up elephant stocks in the same fashion as the Sultanate rulers. Their stables contained even more elephants than earlier. Jehangir is supposed to have possessed 12 000 elephants and over 40 000 in his empire (Jardine, cited by Olivier 1978a). Digby (1971) concedes that considerable numbers of elephants may have been maintained by the Moghuls throughout their kingdom, but considers such high figures as wild guesses and unlikely in view of the much smaller numbers available during the period of the Delhi Sultanate. Part of the confusion lies in the fact that only a fraction, usually less than one-third, of the total numbers held would be suited for direct use in war. It is also not clear whether the higher figure of 40 000 refers to the total numbers of tame and wild elephants estimated to be present within the kingdom. Nevertheless, it is not an impossible figure even if this refers only to tame elephants. Over-exploitation of the wild population, combined with the high longevity of elephants, could have resulted in a large domestic stock for at least a short period. Statistics of elephant captures during the nineteenth and twentieth centuries support such a possibility.

Whatever the true numbers of elephants held by the Moghuls, it is certain that the wild populations of central India, from Dohad (74° E) eastward to Mandla (81° E) were for all practical purposes wiped out (see Fig. 2.2). The distribution of elephants during the late Moghul period seems to have persisted into the nineteenth century. Jerdon (1874) gives its distribution as the Himalayan foothills from Bhutan westwards to Dehra Doon, and in central India from Midnapore (Bengal) to Mandla (Madhya Pradesh) and south nearly to the Godavary. He also mentions that elephants had recently disappeared from the Rajmahal hills. At least three distinct elephant regions had emerged: those in southern India, in central India and a long narrow belt along the Himalayan foothills from Dehra Doon in the northwest extending into northeast India. Each region had further fragmented populations. In the south, certain hills of the Eastern Ghats such as the Shevaroys and the Kollimalais still held small elephant herds. Francis (1906) mentions that elephants were also seen in the Kalrayans during the mid-nineteenth century, though they had disappeared by his time. By the late nineteenth century there were many isolated herds in central and northern India (Imperial Gazetteer of India 1907). References to stray herds mention Nahan and Ambala (Punjab), Udaipur state (Rajasthan), Bilaspur district (Madhya Pradesh) and Parkal (Andhra Pradesh). These highlight the process of habitat fragmentation, isolation and the ultimate disappearance of elephants.

Another element accelerated the decline of the elephant: sport hunting of elephants introduced by the British, probably during the seventeenth or eighteenth century. The *shikar* literature of the nineteenth century testifies to the slaughter of elephants in the Indian sub-continent, Sri Lanka and Burma. Prior to 1873 the government offered rewards for the killing of elephants, both male and female and of all sizes, in order to control damage to crops. Fletcher (1911) records that one man alone killed about 300 elephants, mostly cows and calves, in the Wynad of southern India. In neighbouring Sri Lanka about 3500 elephants were shot in the Northern Province in three years up to 1848, and 2000 were shot in the Southern Province during 1851–55. One man named Rogers alone killed 1300 of these elephants (references are given in Gooneratne 1967 and Olivier 1978a). In October 1873 the Elephant Preservation Act came into force in the Madras Presidency. However, elephants could still be killed on private land. The British also began cultivating tea and coffee on a large scale in the hills of northeastern India and the Western Ghats of southern India during the nineteenth century. Many planters were also hunters.

The twentieth century has witnessed an increase in the human population in India from 236 million in 1901 to an estimated 790 million in 1988. The opening of new land for agriculture along the Himalayan foothills has separated the northwestern and northeastern elephant populations. Since independence in 1947 the accelerated development of industry, hydroelectric and irrigation dams, mining and agriculture has further reduced the elephant's habitat. Once the forests of the Western Ghats and the Terais were rid of malaria, they became home for immigrants from the plains. The last strongholds of the elephant had been conquered by man.

The story in other Asian countries has many similarities and some differences with the Indian situation. These are summarized by Olivier (1978a) and will not be repeated here. In China a major difference was that elephants did not have any religious significance and were usually exterminated as vermin. Elephants enjoyed a close cultural and religious association with man in Sri Lanka, Burma, Thailand and Kampuchea. In Malaysia and Sumatra a former cultural association with elephants has died.

The twentieth century elephant still finds a place in human affairs in Asia. It greets visitors to temples and carries tourists inside wildlife sanctuaries in India. Burma's force of 5400 domestic elephants is vital to its logging industry. During the Second World War (1939–45) the elephant was prized by both the British and the Japanese in their battle for Burma. While armoured tanks were firing in Europe, these skilled 'sappers' were patiently constructing roads and bridges, and transporting troops and supplies across treacherous terrain in Burma (Williams 1950). More recently, during the

Vietnam War, elephants were bombed by American planes to prevent the Vietcong from using them as transport. In a way, the elephant has not yet died as an instrument of war.

1.3 Elephants captured during the nineteenth and twentieth centuries

The number of elephants captured for domestication or eliminated through shooting during the nineteenth and twentieth centuries in Asia far exceeds the present wild population (Table 1.1). Between 1868 and 1980 the available figures of captures in the Indian sub-continent add up to 19 000 elephants, assuming that an annual average of 400 elephants were captured in Assam for ten years during the late nineteenth century. Considering the lacunae in data, it is estimated that between 30 000 and 50 000 elephants would have been captured or killed in control measures during this period. Whatever the true numbers, the offtake has been consistently very much higher from northeastern India than from southern India. This is due to differences in the traditional capture methods. Entire herds were taken in the north but only solitary animals in the south, with the exception of the *kheddah* captures in the southern Mysore state, introduced by G. P. Sanderson in 1874.

Elephant populations in Sri Lanka were also seriously depleted through capture and slaughter during the nineteenth century. Nearly 17 000 elephants were captured during 1911–82 in Burma. The steady decline in the average annual numbers captured in Burma indicates a decline in wild populations. The available figures suggest that over 100 000 elephants have been captured in the whole of Asia during the past century. After the elephant was domesticated about four thousand years ago, anywhere between two and four million elephants may have been captured by man.

I have gone through this short exercise in history and numbers not to establish exact figures but merely to give an idea of the intensive interaction between the Asian elephant and people, resulting in enormous depletion in habitat and elephants. The magnitude of decline has been far greater for *Elephas maximus* than for *Loxodonta africana*. The Asian elephant has declined primarily because of reduction in habitat and captures, whereas the African elephant has been subject more to hunting pressure. About one million *Loxodonta* still inhabit a relatively large proportion of the African continent, although their numbers are also declining (Douglas-Hamilton 1987), whereas less than 50 000 *Elephas* are now confined to a small fraction of their original range.

Table 1.1. *Partial record of elephants captured or killed in Asia during the nineteenth and twentieth centuries*

Region	Number of elephants captured/killed	Period of capture	Number of years	Average number per year	Remarks and source of information
Indian sub-continent					
Northeastern region					
Dacca hills	413	1868–76	7	59	Sanderson (1878)
Chittagong	85	1875–76	1	85	Sanderson (1878)
Dacca hills	503	1876–80	3	168	Balfour (1885)
Assam	c. 400	late 19th century	1	400	Annual catch in Garo hills for at least 10 years. Imperial Gazetteer (1907)
Lakhimpur, Assam	39	1904	1	39	Imperial Gazetteer (1907)
Garo hills, Assam	255	1911–14	3	85	Stracey (1963)
Goalpara, Bengal	621	1916–17	2	311	Stracey (1963)
Sibsagar–Naga hills	900	1934–36	3	300	Stracey (1963)
Assam	3026	1937–50	12	252	Figure supplied by E. P. Gee to Deraniyagala (1955)
Assam	1200	1955–60	6	200	Stracey (1963)
All northeast states	5564 +	1961–80	20	278	Lahiri Choudhury (1986)
	586	1961–80	20	29	Declared as rogues and shot
Southern India					
Mysore	718 +	1874–99	25	29	Kheddah captures, Neginhal (1974)
Mysore	1119	1904–71	69	16	Kheddah captures, Neginhal (1974)
Cochin	28	1902–04	3	9	Imperial Gazetteer (1907)
Mudumalai	130	1910–26	17	8	Forest department records
Madras Presidency and Tamilnadu	525 +	1926–80	55	10	Captured in pits, compiled from Forest Department records
Burma					
	8340	1910–30	20	417	Blower (1985), based on other
	3740	1930–50	20	187	references. See also
	2940	1950–70	20	147	references in Olivier (1978a)
	1560	1970–82	12	130	
	3370	1928–41	13	259	Elephants destroyed. References in Olivier (1978a)
Malaysia					
Perak	88	1948–69	22	4	All references for Malaysia in
Entire country	174	1960–68	9	19	Olivier (1978a). All records
	36	1970–76	7	5	of elephants shot or poisoned
Sri Lanka					
Matara	149	1829	1	149	Captured
Matara	370	1850	1	370	Captured
Northern Province	c. 3500	1845–48	3	1167	Shot
Southern Province	c. 2000	1851–55	4	500	Shot. One person named Rogers alone shot 1300 elephants. References in Gooneratne (1967) and Olivier (1978a)

2

Status and distribution of the Asian elephant

The first comprehensive review of the status and distribution of the Asian elephant was that of Olivier (1978a), made mainly on the basis of existing literature and a questionnaire survey. Since then the efforts of the Asian Elephant Specialist Group of the International Union for Conservation of Nature and Natural Resources, with the help of the World Wide Fund for Nature, have produced a more accurate picture, especially in surveyed areas of the Indian sub-continent, Sri Lanka, Thailand and Sumatra. These surveys have generally revealed that more Asian elephants exist than was hitherto believed. However, vast areas of the elephant range in Southeast Asia have yet to be systematically classified. Estimates of elephant numbers, even for surveyed regions, are mostly educated guesses. The logistics of systematically covering the densely forested regions of Asia are overwhelming. Unfavourable political conditions prevent field work in many countries.

This chapter provides an update of the status and distribution of the Asian elephant and outlines the major conservation issues. Most of the information summarized here is available only in unpublished reports of projects sponsored by the Asian Elephant Specialist Group of IUCN and by the WWF (Santiapillai 1987). The distribution of elephant populations often cuts across political boundaries over the continental mainland and also in certain islands of South and Southeast Asia.

Elephants occur in the following regions and countries.

(a) Indian sub-continent: India, Nepal, Bhutan and Bangladesh.
(b) Continental Southeast Asia: China, Burma, Thailand, Kampuchea, Laos, Vietnam and Malaysia.
(c) Island Asia: Andaman Islands (India), Sri Lanka, Sumatra (Indonesia) and Borneo (Malaysia and Indonesia).

2.1 Indian sub-continent

The distribution of elephants in the Indian sub-continent can be considered under four widely separated regions (Figs. 2.1 and 2.2). Within any one region there are further isolated elephant habitats.

2.1.1 *Southern India*

The elephant is distributed over forested hilly tracts of the Western Ghats and adjacent Eastern Ghats in the southern states of Karnataka, Kerala and Tamilnadu. Its range lies between $8°15'$ N–$15°30'$ N and $74°15'$ E–$78°$ E. The elephant's habitat encompasses a diversity of vegetation types from dry scrub through deciduous forest to wet evergreen forest, with a corresponding variation in annual rainfall from 50 to 500 cm. Elephants are found at altitudes from 100 m to over 2000 m above mean sea level (msl). The following account of status and distribution under 10 sub-regions is based on Nair & Gadgil (1978), Nair, Sukumar & Gadgil (1980) and Sukumar (1985, 1986a).

(a) North Kanara

The North Kanara district of Karnataka is the northern limit of elephant distribution in southern India. The elephants are scattered at very low densities. There has been considerable incursion into the forests by cultivation, mining and a series of reservoirs under the giant Kalinadi hydroelectric project. Large areas of forest, classified as 'minor forest' to cater for the needs of the people, are now in a highly degraded state. In spite of the extensive forest area remaining, the Dandeli Wildlife Sanctuary alone covering 5729 km^2, the elephant population has declined sharply owing to persecution by man. Only a few scattered herds, comprising about 40 individuals, survive in the region.

(b) Crestline of Karnataka Western Ghats

To the south of North Kanara, the montane evergreen forests and grasslands along the crestline of the Western Ghats extend as a narrow belt through the Mudigere, Pushpagiri and Brahmagiri hills. Elephants occur at a very low density in this 1900 km^2 tract. Perhaps fewer than 60 elephants in at least five populations are found here.

(c) Malnad plateau – Bhadra

The Malnad plateau lies to the east of the crestline, separated from it by a wide belt of coffee plantations and cultivation. Vegetation is mainly deciduous forest. An irrigation project on the Tunga and Bhadra rivers is

12 *Status and distribution*

situated here. The elephant habitat includes the Bhadra and Shettihally Wildlife Sanctuaries (area 827 km^2). Some of the elephant herds have been isolated by reservoirs and cultivation. A tentative estimate of 100–150 elephants can be given for this region, based on the Forest Department census figures.

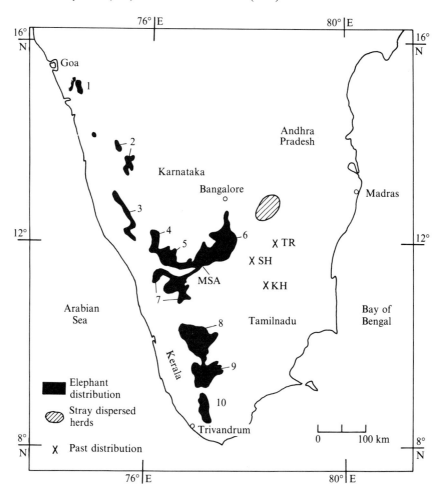

Fig. 2.1. Map of elephant distribution in southern India. The major regions described in the text are: 1, North Kanara; 2, Malnad-Bhadra; 3, crestline of Western Ghats; 4, Nagarhole; 5, Bandipur–Nilgiris; 6, Eastern Ghats; 7, Nilambur–Attapadi; 8, Nelliampathis–Anamalais; 9, Periyar; 10, Agasthyamalai. The location of the main study area (MSA) in the Eastern Ghats is shown. The hatched area shows the range of a few herds which have dispersed there from the Eastern Ghats since 1984. References to past distribution include Tiruvanamalai (TR), Shevaroy hills (SH) and Kollimalai hills (KH).

(d) Nagarhole–Kakankote–North Wynad

The deciduous forests of Kerala North Wynad, Nagarhole National Park and Kakankote (total area 1250 km^2) extend from south of the Cauvery river to the Kabbini river. Extensive teak plantations are seen at Nagarhole. The perennial Lakshmanthirtha and Nagarhole rivers flow through this relatively high-density elephant region. An earlier estimate of 300 elephants (Nair *et al.* 1980) can now be updated to 600–800 elephants.

(e) Bandipur–Mudumalai–South Wynad–North and East Nilgiris

Although the Kabbini reservoir has submerged a considerable area between Kakankote and the Begur range of Bandipur, elephants still move across a narrow 6 km corridor. The deciduous forests extending from the Kabbini river southwards to the slopes of the Nilgiris constitute one of the finest elephant habitats in southern India. The perennial Moyar river is an important water source. This region includes the Bandipur National Park (874 km^2), the Mudumalai Wildlife Sanctuary (321 km^2), the Kerala South Wynad Sanctuary (251 km^2), the semi-arid Sigur plateau and the northern and eastern slopes of the Nilgiris (700 km^2). The density of elephants in Bandipur–Mudumalai is one of the highest in India and, perhaps, in the whole of Asia. Between 1200 and 1500 elephants may be present in the entire region.

(f) Eastern Ghats (south)

The elephant range in the Eastern Ghats, extending over the states of Karnataka and Tamilnadu, is contiguous with that in the Nilgiris to the southwest. The forested area of nearly 7000 km^2 is hilly; the altitude varies from 250 to 1800 m. Vegetation is largely dry deciduous and scrub thicket but a stretch of montane evergreen *shola* (grove) forests and grassland is found in the Biligirirangan hills. The elephants of this region may be considered under two sectors: a northern sector centred on the Cauvery river and extending southwards to the Palar river, and a southern sector to the south of the Palar.

The Bannerghatta–Anekal range is a narrow belt of scrub woodland extending south from the suburbs of Bangalore city to the Kanakapura and Satnur ranges bordering the northern bank of the Cauvery. The adjoining forests of Tamilnadu constitute the Hosur and Dharmapuri Divisions. To the south of the Cauvery are the Hanur and Madeshwaramalai ranges. Along the Cauvery there is a 100 km stretch of virtually uninterrupted dry deciduous forest, although owing to the hilly terrain the animals have access to its waters only in certain places. This sector has an estimated 700 elephants. Since 1985 at least three herds, comprising about 30 elephants, have dispersed north into the Chitoor district of Andhra Pradesh.

South of the Palar river, the Bargur hills extend westwards to the Biligirirangan hills. The forest ranges include Bargur, Andhiyur, Satyamangalam, Talamalai, Ramapuram, Kollegal, Biligiri Rangaswamy Temple, Chamarajanagar and Punjur. The major threat to the habitat is from numerous pockets of cultivation. About 1200 elephants inhabit this sector, a portion of which forms the main study area for the detailed investigation on elephant ecology (Sukumar 1985) described in the subsequent chapters of this book. Based on this study an estimate of 1800–2000 elephants can be given for the entire Eastern Ghats.

(g) Nilambur–West and South Nilgiris–Attapadi hills
To the west and south of the Nilgiris are the well preserved wet evergreen forests, montane evergreen *shola* forests, grasslands and semi-evergreen forests of Nilambur, New Amarambalam, Upper Bhavani, Kundah, Silent Valley and Attapadi. In the Attapadi hills, where much of the land is under cultivation, there is an entire spectrum in vegetation from wet evergreen in the west to scrub in the east. The dry forests continue along the southeastern slopes of the Nilgiris in Coimbatore Division, through which flows the Bhavani river. South of Attapadi, the forested hills end at the Palghat gap. This region has been recently separated from the region to the north of the Nilgiris by tea plantations at Gudalur between Mudumalai and Nilambur to the west, while to the east the habitat constricts to a very narrow corridor along the Mettupalayam–Coonoor highway. Only a few elephants, usually lone bulls, may still move between these two regions. Between 300 and 500 elephants may be found at a relatively low density in this 1700 km^2 area.

(h) Nelliampathi–Anamalai–Palani hills
South of the Palghat gap, the Nelliampathi, Anamalai and Palani hills form a contiguous elephant habitat. This includes the Parambikulam (270 km^2) and the Anamalai (958 km^2) Sanctuaries. An entire spectrum of vegetation types is available here. In the Anamalais the habitat has been disturbed by a series of hydroelectric projects and their associated canals, impeding the free movement of elephants. Elephants are absent towards the eastern portion of the Palani hills which is largely under settlement. The census figures of the Forest Department indicate that 800–1000 elephants are present in the region.

(i) Periyar–Elamalai–Varushanad hills
From the High Ranges at the southern end of the Anamalais, the Periyar plateau extends southwards up to the Shencottah gap. The reservoir on the

Periyar river is the nucleus of the Periyar Tiger Reserve (777 km^2), a well-known sanctuary for tourists. Vegetation on the plateau varies from evergreen to moist deciduous; dry forest is seen on the eastern slopes of the Srivilliputhur range. Based on the estimate by Vijayan (1980) for Periyar, about 700–900 elephants seem to inhabit the entire region.

(j) Agasthyamalai–Ashambu hills
The Shencottah pass maintains a tenuous link between the Periyar plateau and Agasthyamalai, but elephants do not move across the railway line and the highway in this corridor. South of the pass the elephant habitat includes a portion of the Neyyar (128 km^2), Mundanthurai (520 km^2) and Kalakkad (224 km^2) Wildlife Sanctuaries. Hydro-electric and irrigation dams, rubber and tea plantations are the main disturbances. Since elephants are largely confined to the interior evergreen and semi-evergreen forests and grasslands, their status is not very clear. A tentative estimate is 150–200 elephants.

2.1.2   *Central India*
This region has been surveyed by the Central India Task Force of the Asian Elephant Specialist Group (Shahi 1980; Shahi & Chowdhury 1986). Elephants are found in dry deciduous, moist deciduous and semi-evergreen forests of the Eastern Ghats in Orissa and Bihar states. In Orissa a number of isolated populations totalling 1300 elephants range over 12 000 km^2 within 22 Forest Divisions. Of these only about 375 elephants in the Simlipal Tiger Reserve (2750 km^2) and 300 elephants in the Satkosia Gorge Sanctuary (1274 km^2) have secure habitats. The remaining habitats are deteriorating owing to felling of trees and shifting cultivation. Unless this trend is halted the future of over 600 elephants in small scattered populations is bleak. Apart from a shrinking habitat, another conservation problem is the seasonal hunting of mammals by tribes.

In Bihar at least 335 elephants are distributed as three distinct populations. These include 65 elephants within a 1000 km^2 home range in Palamau Tiger Reserve, 200 in the Singhbhum tract (2250 km^2) and 70 in the Dalbhum tract (*c.* 250 km^2). In 1987 the elephants of Dalbhum had moved into the Midnapore district of adjoining West Bengal state. Asia's largest single-point deposit of iron ore is found in the Singhbhum forests. The future of this habitat depends on how mining develops in the region. A further threat arises from the Forest Department's policy of replacing natural sal (*Shorea robusta*) forests with teak (*Tectona grandis*) plantations. Tribal people agitating against this policy have already destroyed 100 km^2 of forest.

2.1.3 Northwestern India and Nepal

Along the Himalayan foothills in the Terai forest region of Uttar Pradesh, an isolated population of about 500 elephants ranges over the Forest Divisions of West and East Dehradun, Shivalik, Lansdowne, Bijnor, Kalagarh, Corbett National Park (521 km^2), Ramnagar, Tarai Bhabar and Haldwani (Singh 1978). Some of these elephant herds also move into Nepal. Another herd of 25 elephants, which may be isolated, has been reported for Dudhwa National Park (490 km^2). Threats to the habitat from the Ramganga reservoir, the Rishikesh–Chilla power channel and a proposed paper mill have been discussed by Singh (1978).

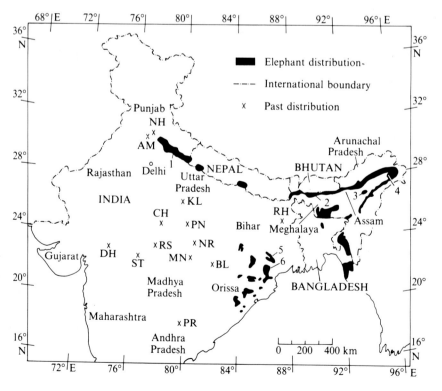

Fig. 2.2. Map of elephant distribution in the northern Indian sub-continent. Some of the important protected areas are: 1, Corbett; 2, Manas; 3, Kaziranga; 4, Namdapha; 5, Dalma; 6, Simplipal. References to past distribution include Nahan (NH), Ambala (AM), Kalpi (KL), Chanderi (CH), Panna (PN), Dohad (DH), Satwas (ST), Raisen (RS), Narwar (NR), Mandla (MN), Bilaspur (BL), Rajmahal hills (RH) and Parkal (PR). Based on Singh (1978), Lahiri Choudhury (1986) and Shahi & Chowdhury (1986).

2.1.4 *Northeastern India and Bhutan*

The elephant habitats in this region have been mapped by the Northeast India Task Force of the IUCN/Asian Elephant Specialist Group in collaboration with the Forest Departments (Lahiri Choudhury 1980, 1986). It was formerly thought that the northeastern region had a contiguous elephant range which also extended into Burma. Now it is known that only a series of fragmented elephant habitats exist, although it seems that solitary male elephants may still move across occasionally. These extend from the Himalayan foothills of Bhutan and northern West Bengal eastwards into the states of Assam, Arunachal Pradesh, Nagaland, Manipur, Mizoram, Tripura and Meghalaya. Some of these populations also extend into Bangladesh and Burma. The vegetation includes swampy grasslands along the floodplains of rivers such as the Brahmaputra, moist deciduous forest and evergreen forest on the Himalayan slopes.

(a) *North Bengal – North Assam – Arunachal*

One of the largest elephant populations in Asia extends along the Himalayan foothills and plains from northern West Bengal eastwards through Assam, Bhutan and Arunachal Pradesh. In West Bengal, out of an estimated 155 elephants about 80 elephants range to the west of the Torsa river, while the rest are found to the east of the Torsa and are linked to Bhutan and Assam. In Arunachal Pradesh alone about 10000 km^2 of hilly habitat may be available for elephants. The Forest Department of Arunachal has estimated 2000–4300 elephants for the state. One of the key areas for conservation in Arunachal is the Namdapha National Park (1808 km^2). In Assam the Forest Department has estimated 1200 elephants for the Manas Tiger Reserve (2840 km^2) and 400 for the Darrang West and East Forest Divisions. The elephants to the north of the Brahmaputra river are linked with those to its south through the Tirap district and extend into Nagaland.

(b) *South of Brahmaputra*

A number of isolated populations are also found to the south of the Brahmaputra. One such population is found in the Dibang–Tirap (estimate included with Arunachal) and Dibrugarh region (200 elephants). Another major population of about 1900 elephants inhabits the Kaziranga National Park (696 km^2), Sibsagar and Nagaland hills. A large population in the Garo hills and Khasi hills of Meghalaya, estimated at 2500–3500 elephants, is certainly isolated from the others in the northeastern region. Smaller populations have been described for the Jainti hills and North Cachar region

(150–175 elephants), South Cachar (100–150 elephants) and Tripura (120–150). The picture in the states of Manipur and Mizoram is not clear but the numbers are certainly very low. Apart from the tremendous pressure on the habitat by people and a history of capturing elephants in large numbers, the prospects for conservation of the elephant in northeastern India are affected by a volatile socio-political situation.

2.1.5 Bangladesh

Bangladesh has an estimated 200–350 elephants, of which about 30% may move into adjacent Burma and India (Khan 1980; Gittins & Akonda 1982). Most of the resident elephants are found in isolated populations in the Chittagong hills. Protected areas inhabited by elephants are the Himchari National Park (23 km^2) and the Mainimukh (256 km^2) and Pablakhali (420 km^2) Wildlife Sanctuaries.

2.2 Continental Southeast Asia
2.2.1 China

In China elephants are found only in the extreme south of the Yunnan province bordering Burma and Laos. Up to 230 elephants may survive in the mixed forests of Xishuangbanna (2400 km^2) and Nangunhe Nature Reserves.

2.2.2 Burma

There has been no systematic survey of elephant status and distribution in Burma. Large expanses of forest cover about 350 000 km^2 of the country. Elephants are still widely distributed in Burma, mainly in forested hill tracts, being absent only from the high mountains to the north and in the dry central zone (Fig. 2.3). They are generally more common in the north than in the south. Elephant habitats cover evergreen, semi-evergreen, moist deciduous and bamboo forests. The main elephant centres are the Myitkyina, Bhamo, West Katha and Upper Chindwin Forest Divisions in the north, the western slopes of Arakan Yoma in the west, the Pegu Yoma in Lower Burma and Tenasserim in the south (Blower 1985, based on UNDP/FAO project). The Shan States and Chin hills have very low elephant densities. Five temporary sanctuaries, which do not as yet enjoy legal status, are exempt from capturing operations by the Forest Department's Elephant Control Scheme. Elephants are found in the established Pidaung (104 km^2), Shwe-U-Daung (326 km^2) and Tamanthi (2150 km^2) Wildlife Sanctuaries falling under these exempted regions.

Fig. 2.3. Map of elephant distribution in continental Southeast Asia. Only approximate distributional ranges are shown. Based on Blower (1985), Dobias (1985) and Santiapillai (1987).

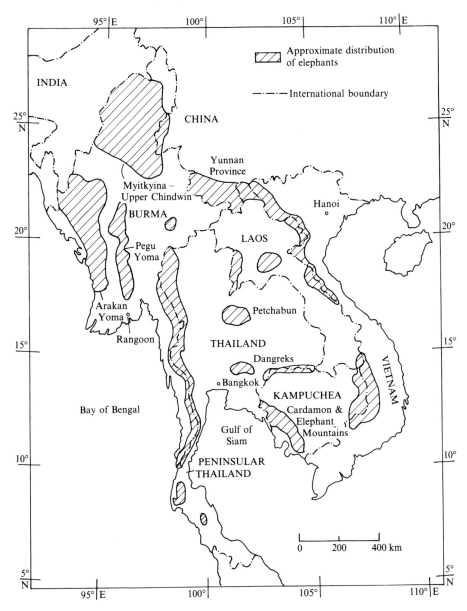

The Forest Department captures an average of 130 elephants every year. Hunting for tusks and meat is widespread. Loss of habitat does not seem to be a major problem at present, although in coastal Arakan and Tenasserim the spread of agriculture has interfered with former seasonal movements (Blower 1985).

During the past five decades, guess estimates fluctuating between 3000 and 10 000 wild elephants have been made for Burma, with the majority of recent estimates falling between 3000 and 7000 elephants. A 1982 report of the Ministry of Agriculture and Forests gave a tally of 6560 elephants, including two large populations of 1637 elephants in Myitkynia and Bhamo of the northern hill ranges and 1812 elephants in the Irrawaddy and Chindwin valleys. G. Caughley (unpublished report) made an even lower estimate of only 3000 elephants for Burma. Going by previous experiences in estimating elephant numbers, I personally consider all these to be underestimates. All the estimates are purely speculative and not based on any objective data for even one region. Historically, Burma's forests have always supported a large offtake of elephants (incidentally, there are 5400 domestic elephants in the country). The teak and bamboo forests of Burma would potentially have a high carrying capacity for elephants. Even if elephants were to exist at a very low density, say 0.1 elephant/km^2, over a 100 000 km^2 forested area, about 10 000 of them may still survive in Burma. The actual numbers may be even higher depending on the area of elephant habitat and elephant densities.

2.2.3 Thailand

Surveys of Thailand's elephant habitats have begun only recently and, thus, available information is preliminary or guess (Lekagul & McNeely 1977). Elephants are distributed rather patchily in small populations over the remaining forested hill tracts of four regions in the country (Fig. 2.3). These regions have a number of protected area complexes, comprising National Parks (NP) and Wildlife Sanctuaries (WS), which are important elephant habitats (Dobias 1985, 1987).

(a) North and west Thailand
A large proportion of Thailand's elephants are found in the Tenasserim range to the north and west along the border with Burma, especially from Mae Hong Son southward to Chumpon. Although a diversity of vegetation types, including evergreen forests, is present, the drier types such as dry deciduous fire-climax savanna woodlands, grasslands and bamboo forests dominate. There is considerable movement of elephants between Thailand and adjoining areas of Burma. One protected area complex in the north of the range comprises the Om Koi WS, Maetuen WS and Mai Ping NP (total area

3400 km²) holding at least 125–175 elephants. A second complex (6913 km²) in the west along the border with Burma includes Huai Kha Khaeng WS, Thung Yai WS, Sri Nakarin NP and Erawan Falls NP with an estimated 300 or more elephants. The Chao Nen dam across the river Kwai and road development seem to have stopped elephant movement between this complex and the Sai Yok NP and Salak Phra WS. For the entire region an estimate of 1300–2100 elephants was made by Lekagul & McNeely (1977).

(b) Petchabun Range

In the Petchabun mountains to the northeast the vegetation includes hill evergreen, dry evergreen, coniferous, mixed deciduous and dry dipterocarp forests. About 125–175 elephants occur in the Phu Luang WS (848 km²). To its south the Phu Kradung NP, Nam Nao NP and Phu Khieo WS complex (2870 km²) holds 250–300 elephants. Along the highway from Chumphae to Lomsaki through the Nam Nao National Park there is considerable encroachment by squatters. Spread of agriculture may further sever habitat integrity.

(c) Dangrek Range

The Dangrek mountains bordering Laos and Kampuchea may hold 100–200 elephants. Hunting of elephants seems common here. Further into the interior of Thailand the Khao Yai NP, Thap Lan NP and Pang Sida NP complex (5252 km²) is better protected. About 78% or 1700 km² of the Khao Yai NP area is available for elephants. A density of 0.15 elephant/km² or 225 elephants for the park was estimated on the basis of dung counts (Dobias 1985). These elephants are separated from those in Thap Lan and Pang Sida by a highway and human settlements.

(d) Peninsular Thailand

A number of populations numbering 900–1500 elephants are scattered between Ranong and Trang in the peninsula. Of these more than 200 elephants occur in the Maenam Pha Chi WS and Kaeng Krachan NP complex (2967 km²) in the north of the peninsula. Some of these may be migrants from Burma. Further south the Khlong Naka WS, Khlong Saeng WS and Khao Sok NP complex (2281 km²) hold more than 100 elephants.

The future of elephants in Thailand may be largely tied up with the system of protected areas and their management. Of the total estimate of 3000–4500 elephants for Thailand, the 29 protected areas covering 26000 km² hold 1300–1700 elephants. However, the populations are largely scattered. Only 13 protected areas are thought to contain numbers in excess of 25 individuals and only 7 areas may have populations above 100 elephants.

2.2.4 Laos, Kampuchea and Vietnam

No worthwhile information is available for these countries (Olivier 1978a). In Laos elephants are still widely distributed in the south (Sayer 1983). They occur in most of the forested areas of Vientiane province and also in the Sayabouri province to the west of the Mekong river (Fig. 2.3). In the northern highland rain forests, the home of hunters and gatherers, they seem to have been largely exterminated except in the Nam Ngum basin, possibly overlapping with the Yunnan elephants of southern China in the upper Mekong basin. About 74 000 km^2 or 40% of Kampuchea is forested. Elephants are found on the border with Vietnam and in the Dangrek range adjoining Thailand up to the Laos border. The Cardamon and Elephant mountains near the southern coast, covered with rain forest, are also potential elephant habitats. In Vietnam the forests near the Laos and Kampuchean borders, especially Buon Don, are refuges of elephants. Olivier (1978a) made a guess estimate of 3500–5000 wild elephants for these three countries. A more recent questionnaire survey gave a tally of 2000 elephants for Laos, 2000 for Kampuchea and 1000 for Vietnam (Santiapillai 1987). With the recent wars and political upheavals in the region it is unlikely that a clear picture will emerge in the near future.

2.2.5 Malaysia

Elephants are widely distributed over peninsular Malaysia from the state of Perlis in the north to Johore in the south and from Selangor in the west to Trengganu in the east, though they occur as small scattered herds (Fig. 2.4). Typically, the elephant habitat is equatorial rain forest, both primary and secondary types. Elephant density is usually higher in secondary habitat than in the primary forest (Olivier 1978b). Estimates of population numbers have varied widely from about 600 to 6000 elephants. The lower numbers, usually derived from a registration of herds known to make contact with human habitation, are certainly underestimates; the higher figures, extrapolated from density estimates, may be rather optimistic (Olivier 1978a). The latest tally of about 800 elephants, admitted to be around 20% underestimate, has been derived from a registration of known herds in the following states (Khan 1985).

(a) Perlis: a single herd of 5 elephants and an adult bull.
(b) Kedah: a number of bulls and herds totalling 44 elephants.
(c) Perak: eleven herds comprising 126 elephants.
(d) Selangor: two herds totalling 6 elephants.
(e) Negeri Sembilan: a total of 13 elephants.

Fig. 2.4. Map of elephant distribution in peninsular Malaysia. The solid circles refer to registered herds outside the main distributional range. Based on Khan (1985) and Santiapillai (1987).

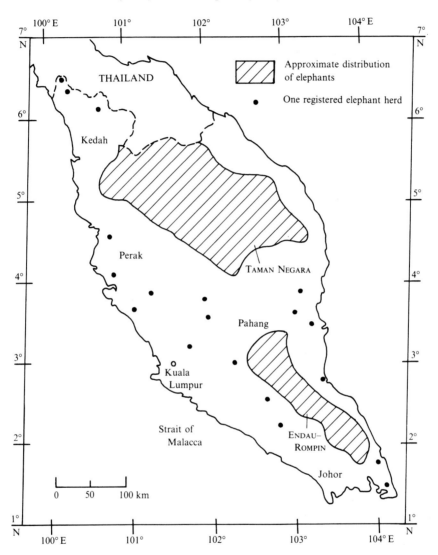

(f) Pahang: fifteen herds totalling 212 elephants are found outside the Taman Negara National Park.
(g) Kelantan: about 30 groups comprising 92 elephants.
(h) Johore: about 94 elephants.
(i) Trengganu: six known herds with 37 elephants.
(j) Taman Negara National Park (4350 km^2): a total of 166 elephants has been estimated for this park, which covers a portion of the states of Pahang and Kelantan.

The low density of less than 0.1 elephant/km^2 and the scattered nature of herds even in contiguous areas may reduce the long-term viability of populations in rain forests. Added to this, the rate of clearing of forest land for agricultural development has been extremely rapid in peninsular Malaysia. This would hasten fragmentation of habitat and isolation of elephant herds. As in many other Asian countries the elephant in Malaysia may survive only in conservation areas such as the Taman Negara National Park and the proposed Endau–Rompin National Park (635 km^2).

2.3 Island Asia
2.3.1 Sri Lanka

Elephants are distributed over a wide area extending from the northwest through the dry eastern zone to the southeast in Sri Lanka (Fig. 2.5). Vegetation in the eastern belt is dry deciduous woodland, scrub, grassland and marshes called *villus* in the flood plains of rivers. Incursion of agriculture, including shifting cultivation, has caused fragmentation of the habitat in many areas (Olivier 1978*a*). A number of attempts have been made to capture and relocate pocketed elephant herds. One successful operation was the translocation of 160 elephants from System H of the Mahaweli Development Scheme to the Wilpattu National Park in 1978–79. Elephant distribution can be considered under four regions.

(a) Northwest region
The Wilpattu NP (1050 km^2) and surrounding areas may have a total of 200 elephants including those translocated from the Mahaweli area. Elephants also occur to the south of Wilpattu in the Mi Oya and Kala Oya basins (Hoffmann 1978). It is not clear whether the elephants in the north-west are isolated from the rest of the population.

(b) Northern province

North of Trincomalee (8°40′ N) the habitat has not been surveyed and its status is not clear. McKay (1973) made a guess estimate of 200–500 elephants for this zone.

(c) Mahaweli Ganga basin

South of Trincomalee is the Mahaweli Ganga basin. The accelerated Mahaweli Ganga Development Programme involving the construction of a series of dams across the country's largest river and opening of new land for

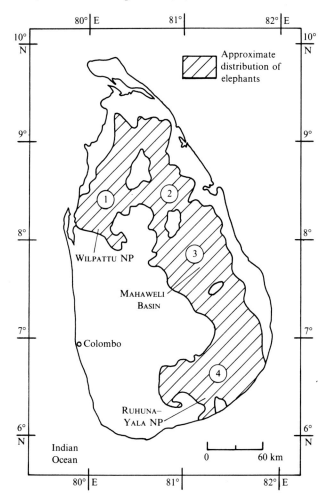

Fig. 2.5. Map of elephant distribution in Sri Lanka. The major regions described in the text are: 1, Northwest; 2, Northern Province; 3, Mahaweli Ganga Basin; 4, Southeast. Source: Santiapillai (1987).

agriculture has been causing significant changes to the elephant's habitat (see Chapter 9). A network of protected areas proposed or located here include the Somawathiya National Park (210 km^2), Wasgamuwa NP (338 km^2), Flood Plains NP (174 km^2), Maduru Oya NP (515 km^2), Trikonamadu Nature Reserve (250 km^2), Minneriya–Giritale NR (420 km^2) and the Nilgala Jungle Corridor (280 km^2). In the north the Somawathiya NP is contiguous with the Hurulu Reserve Forest, while in the south the Nilgala Corridor connects with the Gal Oya NP. Cultivation under Scheme C of the Mahaweli programme will eventually sever the habitat unless a corridor is provided. About 800 elephants are estimated to occur in the Mahaweli basin.

(d) Southeastern region

In the southeast of the island, the complex of Gal Oya NP (518 km^2), Amparai Sanctuary (380 km^2), Lahugala Tank and environs, Ruhuna NP (1079 km^2) and Yala Strict Nature Reserve (286 km^2) is an important stronghold of the elephant. The figure of 650–700 given by McKay (1973) for this region seems to be an underestimate (Olivier 1978*a*).

There has been considerable difference of opinion regarding the total number of elephants in Sri Lanka. McKay (1973) estimated a total of 1600–2200 elephants, but Hoffmann (1975) and Olivier (1978*a*) put it at closer to 4000 elephants. Hoffmann (1978) even revised this to 5000 elephants. The latest tally by the Wildlife Department is about 3000 elephants (A. B. Fernando, personal communication). Elephant densities in Sri Lanka are some of the highest in Asia, comparable to those attained in parts of India. The recent civil war in the northeast may have affected conservation prospects.

2.3.2   *Andaman Islands (India)*

Feral descendants of elephants taken to the Andamans for timber operations range over Interview Island. They even swim across the sea at certain points from one island to another. About 20–30 feral elephants may be present. These are increasingly coming into conflict with people. There is a proposal to capture these elephants by chemical immobilization.

2.3.3   *Sumatra (Indonesia)*

Surveys carried out by the World Wide Fund for Nature have revealed that between 2800 and 4800 elephants are present in Sumatra (Blouch & Haryanto 1984; Blouch & Simbolon 1985; Santiapillai & Suprahman 1985; Santiapillai 1987). Elephants are found throughout

Island Asia

Sumatra except in the provinces of Sumatera Utara and Sumatera Barat (Fig. 2.6). However, the elephants occur in a number of isolated populations of which only five seem to consist of over 200 individuals each.

In December 1982 about 232 elephants were driven into the Padang Sugihan Wildlife Reserve in Sumatera Selatan from a human transmigration project to the north. Along with the elephants earlier present here and in the adjacent Lebong Hitam production forest (3000 km^2), a total of 500 individuals, this population seems to offer the best prospects for long-term conservation. Other important populations include one of 200–300 elephants

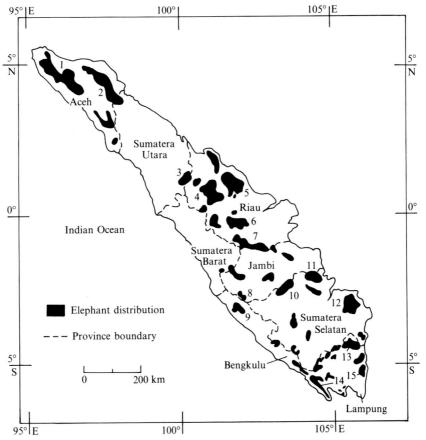

Fig. 2.6. Map of elephant distribution in Sumatra. Of the 44 populations, those believed to have over 100 elephants each are in: 1, Western Aceh; 2, Eastern Aceh; 3, Torgamba; 4, North-central Riau; 5, Siak Kecil; 6, South-central Riau; 7, Southern Riau; 8, Sungai Ipuh; 9, Gunung Sumbing; 10, Air Kepas; 11, Air Medak; 12, Padang Sugihan; 13, Air Mesuji; 14, South Barisan Selatan; 15, Way Kambas. Based on Blouch & Haryanto (1984), Blouch & Simbolon (1985) and Santiapillai (1987).

in the Barisan mountains of western Aceh, 300–450 residing south of the Pase river up to Gunung Leuser National Park in eastern Aceh, 200–300 in north central Riau province and 300–400 in southern Riau. Most of the other scattered populations seem unviable, the majority of them consisting of only 50–100 individuals each. The provinces of Riau, Aceh and Lampung contain most of Sumatra's elephants.

The elephant habitat in Sumatra ranges from lowland rain forest and swamp forest at sea level to montane forest at 2000 m. Although relatively extensive areas of forest still remain, the density of elephants is low, usually below 0.1 elephant/km^2. The highest recorded density of about 0.3 elephant/km^2 in the Padang Sugihan Reserve was attained after the translocation of elephants into the area. The island's forest cover is rapidly shrinking in the face of agricultural expansion, shifting cultivation, mining and transmigration of people from Java. The main nature conservation areas which offer some prospects for the survival of the elephant include the Gunung Leuser National Park (9400 km^2) in southern Aceh, Siak Kecil Reserve (*c.* 1000 km^2) and Bukit Kembang–Bukit Baling Baling Reserve (*c.* 1200 km^2) in Riau, Padang Sugihan Reserve (750 km^2) in Sumatera Selatan, Way Kambas (1235 km^2) and Barisan Selatan (3568 km^2) National Parks in Lampung, and Kerinci Seblat National Park (14 846 km^2) extending over Jambi, Sumatera Selatan and Bengkulu.

2.3.4   *Borneo (Malaysia and Indonesia)*

It is not clear whether the elephants in Borneo are of indigenous origin or have descended from captive elephants presented to the Sultan of Sulu in 1750 by the East India Company and later set free in northern Borneo (Hooijer 1972; Olivier 1978*a*). Today, elephants are restricted to eastern Sabah (a state of the Federation of Malaysia) in northern Borneo and a small contiguous area of East Kalimantan (Indonesia), a total range of 35 000 km^2 (Fig. 2.7).

Two distinct populations with a total of 500–2000 elephants are found in Sabah (Andau & Payne 1985). Of these 100–200 elephants range over the Tabin Wildlife Reserve (1220 km^2) and adjoining areas, mostly logged dipterocarp forest on flat to very steep terrain. The remaining elephants are found within a 16 670 km^2 contiguous area of permanent Reserve Forest in the interior of Sabah. The hilly terrain, from 300–1500 m above sea level, is covered by undisturbed dipterocarp forest with only the peripheral areas being logged. As in peninsular Malaysia the clearing of forest for agriculture is bringing elephants into conflict with people. In east Kalimantan elephants are found only in the upper Sembakung river of Tindung district, particularly in the Ulu Sembakung Reserve (5000 km^2).

2.4 Summary of population estimates

A summary of the elephant population estimates for different countries and the minimum habitat area available is given in Table 2.1. There are 16 745–22 435 elephants within about 85 230 km^2 in the Indian subcontinent, 12 325–22 780 elephants in Continental Southeast Asia within 208 000 km^2 and 5320–10 830 elephants within 143 000 km^2 in the islands of Asia. The total tally is 34 390–56 045 wild Asian elephants over a 436 230 km^2 habitat, of which about 131 820 km^2 (30%) is part of conservation areas such as National Parks and Sanctuaries. The wide margin between minimum and maximum estimates in many regions highlights the difficulties in censusing animals in forest habitats and the lack of systematic surveys for various reasons. Many countries also have large numbers of domestic elephants. A total of about 16 000 captive elephants exist in Burma (5400), Thailand (4800), India (3000), Laos (1300), Vietnam (600), Kampuchea (500), Sri Lanka (500) and Bangladesh (50).

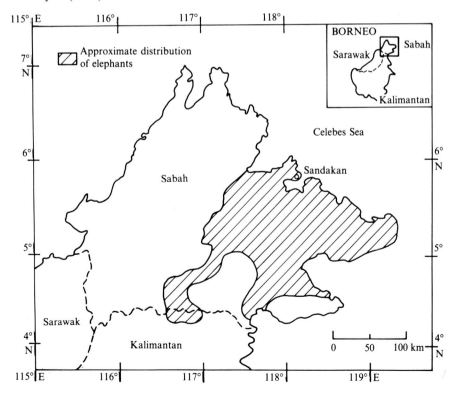

Fig. 2.7. Map of elephant distribution in Borneo. Elephants are found mainly in Sabah and a small area of East Kalimantan. Based on Andau & Payne (1985).

Table 2.1. *Population estimates for the wild Asian elephant*

Region/Country/Locality	Population Min.	Population Max.	Approximate area of elephant habitat (km^2)	Elephant habitat under conservation areas (km^2)
Indian sub-continent				
Southern India				
North Kanara	40	40	>2000	1000
Crestline	60	60	1900	420
Malnad–Bhadra	100	150	>1000	830
Nagarhole	600	800	1500	1250
Bandipur–Nilgiris N. & E.	1200	1500	2200	2200
Nilambur–Nilgiris W. & S.	300	500	2100	2100
Eastern Ghats	1800	2000	7000	1170
Anamalais	800	1000	>2000	1325
Periyar	700	900	>2000	850
Agasthyamalai	150	200	1500	700
Central India				
Orissa	1300	2000	12 000	4415
Bihar	335	335	3500	1170
Northwestern India & Nepal				
Uttar Pradesh	525	525	7580	1600
Nepal	50	85	>2500	1600
Northeastern India & Bhutan				
Bhutan	60	60	2000	1500
West Bengal	155	155	2000	1020
Arunachal Pradesh	2000	4300	>10 000	>3280
Assam & Nagaland	3500	3500	>10 000	>3600
Meghalaya	2750	3825	>8000	0
Tripura	120	150	2900	0
Manipur, Mizoram	?	?	?	?
Bangladesh	200	350	>1550	700
Sub-total	16 745	22 435	85 230	30 730

Summary of population estimates

Table 2.1—continued

Region/Country/Locality	Population Min.	Population Max.	Approximate area of elephant habitat (km²)	Elephant habitat under conservation areas (km²)
Continental S.E. Asia				
Burma	6000	10 000	>100 000	> 6000
China	100	230	>3000	2400
Thailand				
N. & W. Thailand	1300	2100	>17 000	11 000
Petchabun	375	500	> 5000	5000
E. & W. Dangrek – Khao	350	450	>7000	6000
Yai Peninsula	900	1500	>6000	5250
Laos, Kampuchea & Vietnam	3500	5000	>50 000	?
Malaysia (Peninsula)	800	3000	>20 000	5100
Sub-total	12 325	22 780	208 000	40 750
Island Asia				
Sri Lanka	2000	4000	>8000	6120
Andaman Islands	20	30	?	0
Sumatra				
Lampung	550	900		
Bengkulu	100	200		
Jambi	200	500	>100 000	48 000
Sumatera Selatan	250	650		
Riau	1100	1700		
Aceh	600	850		
Borneo	500	2000	35 000	6220
Sub-total	5320	10 830	143 000	60 340
Grand total for Asia	34 390	56 045	436 230	131 820

Sources of information are given in the text.

2.5 Conservation issues

Issues associated with the conservation of elephants in Asia are those concerning man's indirect impact on the elephant through exploitation of its habitat or a direct impact in reducing elephant numbers, and the elephant's impact on human interests. In addition, human social, administrative and political problems may also affect the elephant. These issues as applicable to different regions and countries are outlined here. Their implications, whether negative, neutral or positive, for elephant conservation will be discussed in detail in the following chapters.

2.5.1 *Exploitation of habitat*

On a historical time scale, loss of habitat has been the ultimate cause of decline of elephant populations throughout Asia. This factor is still very serious in many Asian countries. The elephant's habitat either undergoes serious change as a result of the human quest for natural resources or may be lost owing to human occupation. People exploit the elephant's habitat for a variety of plant resources such as fruits, bark, fodder, fuel and timber. In the process the vegetational structure changes from primary climax types to secondary forms. Exploitation of timber may not always be carried out according to ecologically sound prescriptions, resulting in devastation of the natural habitat.

In the Sumatran province of Sumatera Selatan, the construction of roads for oil drilling and extraction in the Subanjeriji area led to the destruction of a 650 km^2 stretch of forest now holding less than 50 elephants (Blouch & Haryanto 1984). Although the Sumatran Forestry Department's regulations limit logging to trees over 50 cm diameter, a survey of stored logs outside saw mills by Nash & Nash (1985) revealed that only 4% of the logs were of legal size.

Conversion of natural forest to commercial timber or pulp-wood plantations is another issue of relevance. Monoculture plantations of eucalypts (*Eucalyptus* spp.), wattle (*Acacia* spp.) teak (*Tectona grandis*) and silver oak (*Grevillea robusta*) occupy about 10% of the elephant habitat in southern India. In Sumatra the Forestry Department's stated goal is to convert 62 000 km^2 of selectively logged production forests into timber plantations by the turn of the century. The impact of such practices on the space and food resources of elephant populations has to be considered.

2.5.2 *Shifting cultivation*

Slash-and-burn shifting agriculture is a traditional practice of people inhabiting tropical moist forests over much of Asia. In the north-eastern Indian states of Meghalaya, Arunachal Pradesh, Manipur, Mizoram,

Nagaland and Tripura, extensive areas of elephant habitat are subject to shifting cultivation (Lahiri Choudhury 1980). With the expanding human population, a reduction in available land area per capita and a shortening time period between successive use of sites, the forest cover may be permanently depleted in areas holding elephants. Between 1950 and 1980 about 1660 km^2 or 25% of Tripura's Reserved Forests were lost through shifting cultivation and refugee settlement. In Nagaland about 5784 km^2 or 35% of the geographical area of the state is currently under shifting cultivation or in fallow; in Meghalaya about 3115 km^2 or 43% of the total Reserve Forest area in the Jainti hills is in a degraded state because of this practice. In the central Indian state of Orissa about 11 000 km^2 or 55% of the area of seven Forest Divisions holding elephants is under shifting cultivation (Shahi & Chowdhury 1986).

Sumatra also faces the problem of shifting agriculturalists, who often move in after logging companies move out of production forests they had been exploiting. The Way Terusan production forest in eastern Lampung, officially listed as 840 km^2, is now highly reduced and fragmented owing to shifting cultivators occupying it after all the merchantable timber had been removed (Blouch & Haryanto 1984). A similar situation prevails in northern Sumatera Selatan. Illegal forest conversion is widespread in the Mount Rindingan region of northern Lampung. Apparently local government officials unlawfully sell the rights to clear protected forests, defeating the efforts of the government resettlement programme which had been removing illegal settlers from the same area. According to one estimate, shifting cultivation is responsible for the loss of 15 000 km^2 of forest every year in Indonesia.

2.5.3 Spread of agriculture

Apart from shifting agriculture in elephant habitat, the establishment of permanent cultivation is continuing at an alarming rate in peninsular Malaysia, Borneo and Sumatra. The Malaysian and Indonesian governments have embarked on a massive expansion of agriculture through clearing of existing natural forest for planting oil palm, rubber, banana, coconut, sugar cane, cocoa and paddy.

In Peninsular Malaysia the operations of the Federal Land Development Agency (FELDA), the largest such agency, covered a total of 5200 km^2 in 1980, with new land being opened up at a rate of 400 km^2 per annum (Blair 1980). About half the elephant range (35 000 km^2 in 1985) in Sabah was allocated for agriculture during the period 1978–84. As clearing of the forest proceeds the elephant range will be split into two portions.

In Indonesia the loss of natural forest cover to agriculture occurs in a different context. Of the total human population of 160 million (in 1985) in Indonesia, the island of Java accounts for 97 million. At the present rate of increase the population is projected to double in 35 years. Due to the enormous pressure on land in Java, the Indonesian government is shifting people from Java to Sumatra under a Transmigration Scheme. Some available figures illustrate the enormous impact of the project on existing elephant forests in Sumatra. In northern Lampung province about 1800 km^2 of forest in the middle of elephant habitat was designated to receive settlers. During 1981–83 about 26 500 families or 107 980 individuals moved into the area. It is planned to settle 90 000 families in transmigration project areas of Jambi, Aceh and Riau between 1984 and 1989. In the Langgam area of Riau about 500 km^2 has been designated to receive 9500 families. To provide employment for the migrants, the area of oil palm plantations in Riau is to be increased from an existing 340 to 4200 km^2. On an average the resettlement of one person means the loss of one to two hectares of forest land. New agricultural projects make more impact on elephant habitat than the statistics reveal. Often the land cleared for cultivation is well-drained flat lowland forest, which is prime habitat for elephants. This is the main threat to elephants in Riau, which is largely flat and hence attractive for transmigration schemes.

2.5.4 Hydro-electric and irrigation dams

Elephants are usually displaced by agriculture from the lowland plains into hilly regions. Hilly terrain is particularly suitable for another form of development: the construction of dams across rivers to store water for generating electricity or irrigating cultivated land in the plains. The Western Ghat hill chain of peninsular India offers the possibility of storing large quantities of water in reservoirs. The steep drop of the hills to the west provides ideal conditions for generating hydro-electric power. The dry tracts to the east of the mountains benefit from irrigation. There are about 40 hydroelectric or irrigation reservoirs ranging in size from 5 to 150 km^2 (submersion area) within or adjoining elephant habitat in southern India.

Dams have made a mixed impact on elephant habitat. On the one hand they have submerged prime river valley habitats, disrupted traditional movement patterns and fragmented the habitat; on the other they have sometimes provided a perennial water source in an otherwise dry habitat for elephants. In the Anamalai hills of the Western Ghats, a series of reservoirs and associated canals under the Parambikulam–Aliyar project has hampered the movement of elephants and even been a direct death trap for them.

Further south, the Periyar Reservoir is the nucleus of an important wildlife preserve.

Under the accelerated Mahaweli Ganga Development Project, the construction of five large dams is either complete or in progress across Sri Lanka's longest river (Alexis 1984). About one million people will be eventually moved into the Mahaweli project area to occupy the newly opened agricultural land. The project has begun to disrupt the movement of elephants, fragment the habitat and displace a considerable number of elephants (Ishwaran & Banda 1982). As yet the construction of dams does not seem to be a major issue for elephant conservation in other Asian countries.

2.5.5   *Capture of elephants*

Since the elephant is still prized as a beast of burden in Asia, both legal and illegal captures continue in many countries. Under an Elephant Control Scheme of the Forest Department in Burma, an average of 130 elephants were captured annually during 1970–82 to cater to the timber industry. Although quantitative data are scanty, it has been suggested that this offtake is higher than the annual rate of growth of the wild elephant population (Blower 1985). Some elephants are also believed to be smuggled alive into Thailand, where they are in demand for logging operations.

Capture of elephants is banned in India, though in 1985 the Assam and Meghalaya Forest Departments were permitted to capture 200 elephants to reduce crop depredation. Illegal capture is rife in the state of Arunachal Pradesh (Lahiri Choudhury 1986). In fact, an illegal trade in elephants seems to cover a wide area in continental Southeast Asia involving northeastern India, Burma, Thailand, Kampuchea, Laos and Vietnam.

2.5.6   *Hunting of elephants*

Elephants are shot mainly for ivory and to a lesser extent for meat. In one sense the Asian elephant is more fortunate than the African elephant because female *Elephas* do not possess tusks. Hence, hunting for ivory is confined to male tuskers. Tuskless males (known by various local names such as *makhna, aliya*, etc.) exist in widely varying frequencies in different regions.

Naturally the male elephant is under more serious threat from ivory poaching in regions where tuskers are more frequent. During 1980–86 at least 100 male elephants were killed every year for their tusks by poachers in southern India. This is one of the most important conservation issues in this region. Ivory poaching has also been reported in the states of Assam, Meghalaya, Nagaland and Mizoram. In the last two mentioned states elephants are also hunted for meat by some tribes. Mortality from hunting for ivory and meat in the Arakan Yoma, Lower Chindwin, Pegu Yoma, Shan States, Tenasserim and Katha District of Burma is reported to exceed the numbers captured annually. Local elephant populations in Tenasserim seem to have been eliminated by poachers (Blower 1985). In Thailand's protected areas at least 91 elephants, mostly tuskers, representing 10% of the total estimated numbers were reported poached during 1975–79 (Dobias 1985). The Moi people of Vietnam are known to consume elephant meat. Illegal hunting also occurs in other countries, although no figures are available. It must also be emphasized that official figures should be taken as the minimum estimates; the actual extent of hunting may be far higher in many regions.

2.5.7 Crop raiding and manslaughter by elephants

When elephants affect people by damaging cultivated crops or resorting to manslaughter, they create powerful social and economic justifications for their elimination. Crop raiding by elephants occurs to varying extents throughout their distributional range, wherever cultivation abuts onto elephant habitat. Depredation seems to be a major problem in parts of Malaysia (Blair, Boon & Noor 1979), Sumatra (Blouch & Haryanto 1984; Blouch & Simbolon 1985; Santiapillai & Suprahman 1984), Sri Lanka (McKay 1973) and India (Mishra 1971; Lahiri Choudhury 1980; Sukumar 1985, 1986a,b, 1989a,b). The problem exists over a wide range of habitat types having low to high elephant densities.

In purely economic terms the value of crops damaged is higher when these are perennial crops such as oil palm, coconut or rubber, compared with seasonal crops like millets, cereals and pulses. Thus, in Malaysia, the loss due to destruction of 4000 hectares of oil palm and rubber trees annually amounted to US$20 million for the Federal Land Development Agency alone (Blair et al. 1979; Blair 1980). Against this there may be fewer than 1000 elephants in the peninsula. On the contrary, a population of about 2500 elephants in the Karnataka state of southern India damages mainly millet crops worth only about $0.15 million annually (Sukumar 1986b).

Depredation inevitably accompanies the spread of agriculture, either permanent settlement or shifting cultivation. Often the problem is due to pocketed herds isolated in small patches of forest amidst a vast sea of cultivation. Adult male elephants are usually the most tenacious raiders, although family herds also cause widespread havoc by coming in large groups. Herds of up to 70 elephants have been known to devastate crop fields and pull down huts in which harvested grains are stored in West Bengal state (Lahiri Choudhury 1980). Crop raiding is a serious conservation problem in the Eastern Ghats region of southern India, owing to the numerous enclaves of cultivation which have made dents on habitat integrity.

Manslaughter may accompany the elephant's incursion into cultivation. Roughly half these cases occur within human settlement. Even some of the killings within the forest may have their origin in settlements where elephants may be injured by bullets and later turn into 'rogues'. From available information it seems that manslaughter by elephants is most serious in India where between 100 and 150 people fall victim annually.

Apart from shooting elephants, people largely depend on primitive and ineffective methods of keeping elephants out of cultivation. Modern devices such as the high-voltage electric fence have been in use only since the late 1970s, notably in Malaysia. It must be admitted that there are no easy methods of keeping elephants out of human settlement.

2.5.8 Political unrest and war

Political upheavals often cause severe damage to conservation prospects. They may result in a breakdown of the administrative machinery in protected areas, cause significant changes to the habitat and even directly cause the death of target species such as the elephant.

Years of political unrest in the Indian states of Mizoram, Nagaland, Manipur and Tripura have meant low priority for conservation by the administration. In Sri Lanka the administration of protected areas suffered after the civil war began in 1983. Both government troops and guerillas overran the forests. Since then no reliable information has been available on the status of the elephant in the country.

In the Vietnam War during the 1960s and 1970s, American planes directly bombed elephants to prevent the Vietcong from using them as transport, while troops shot them for meat (Olivier 1978a). As a result of this war about 20 000 km^2 of forests were contaminated by bombs, defoliants and herbicides and a further 10 000 km^2 were rendered 'unusable' (Vo Quy 1986). The long-term impact of the war on the forests, and ultimately on the elephant, has not been evaluated. The continuing political uncertainty in neighbouring Kampuchea further complicates the problem.

2.5.9 Legal and administrative inadequacies

In most countries the elephant itself enjoys some kind of legal protection but the same is not true of the elephant's habitat. The elephant's range in southern India is almost entirely Reserved Forest and, hence, legally protected. However, under the Indian Constitution the forests in the northeastern states of Arunachal Pradesh, Meghalaya, Nagaland, Tripura,

Manipur and Mizoram do not have the same legal status. Only a minor proportion of the forests, varying from 24% in Arunachal Pradesh to about 9% in Nagaland and Meghalaya, are government Reserved Forests, the rest being village forests, community forests, privately owned forests or unclassed state forests belonging to the district councils. Only the Reserved Forests are managed in any meaningful sense (Lahiri Choudhury 1980). The remaining forests are subject to uncontrolled slash-and-burn shifting cultivation. Since most of the elephants inhabit these non-government forests they ultimately enjoy little legal immunity. Similar problems exist in many other countries. Even in protected areas, the relevant law enforcement agencies are often inadequately staffed to protect both the habitat and the elephant.

Some of these conservation issues were the focus of an ecological study on elephants carried out in southern India (Sukumar 1985) and discussed in the subsequent chapters.

3

The main study area and study methods

The aim of the detailed ecological study was to understand the conflict between elephants and people in relation to the life history of the elephant. The main considerations in selecting the study area were that it should be representative of elephant habitats, harbour a viable and observable elephant population, and exemplify the problem of elephant–human conflict. After an initial survey of elephant habitats in southern India during June–September 1980, a portion of the Eastern Ghats comprising the Biligirirangan hills, the Talamalai plateau and the Moyar river valley was selected as the main study area. This region has a number of cultivated enclaves plagued by elephant–human conflict, encompasses a diversity of vegetation types from dry thorn scrub through deciduous forest to patches of evergreen forest and grassland, and harbours a medium elephant density (about 0.5 elephant/km^2) relative to other populations in southern India. The Mudumalai–Bandipur tract, which has a more abundant (over 1 elephant/km^2) and observable elephant population but is not as suitable for studying elephant–agriculture conflict, was a minor study area.

The main questions the study aimed to answer were the following.

(a) How are the elephants of this tract distributed? How do they utilize the different habitat types seasonally?
(b) What are the plants they consume? What is their foraging strategy?
(c) What is the elephant's impact on the vegetation? Can the primary production sustain the current elephant density?
(d) What are the patterns underlying elephant feeding on cultivated crops? How much nutrition do they derive from crops? Why do they raid crops? What are the economic implications of crop depredation?

40   *Main study area and methods*

(e) What are the circumstances in which people are killed by elephants?
(f) How do people manipulate the elephant's habitat? How does this affect the food resources for elephants?
(g) How many elephants are killed during crop raiding or poached for their tusks? What is the significance of poaching for the ivory trade?
(h) How is the elephant population faring in its interaction with people? Are its numbers increasing, decreasing or stable?
(i) What can be done to minimize elephant–human conflict and conserve elephant populations?

This chapter provides a brief description of the main study area, including its topography, hydrology, climate, land-use, vegetation, mammals and human communities. The study methods are also briefly described.

3.1   The study area

The study area is situated between $11°30'$ N–$12°0'$ N and $76°50'$ E–$77°15'$ E (Fig. 3.1). It comprises a forested area of 1130 km^2 spread over the Chamarajanagar, Kollegal and Satyamangalam Forest Divisions in southern India. The cultivated enclaves occupy an additional 70 km^2 within the study area.

From the Mysore plateau, at an average elevation of 750 m, the Biligirirangan hills rise a further 1000 m. The two central chains of hills, running from north to south, feature a number of peaks above 1500 m, which present a vista of grasslands and evergreen *shola* (grove) vegetation. The enclosed valleys contain moist deciduous forest. Both to the east and to the west, the lower hills and valleys from 1250 m to 750 m become progressively drier away from the central ranges, with dry deciduous forest changing into degraded scrub. To the south the hills slope steeply to the Coimbatore plains and the Moyar river valley (250 m), where a strip of dry thorn forest is found at the foothills. South of the Moyar river rises the Nilgiri mountain range (highest peak 2636 m), which forms part of the Western Ghat chain of peninsular India. Geologically, the entire region is formed by the Archean group of rocks, mainly the gneisses.

3.1.1   *Hydrology*

The region has two drainage systems: a major one emptying into the Cauvery river and a minor one into the Moyar river. Given a general northwards aspect in this area, most of the streams ultimately reach the Cauvery, which flows in an eastward direction to the north of the

Fig. 3.1. Satellite image of the main study area taken by LANDSAT in Band 5 on 21 January 1982. The dark areas are forested and the white areas are under cultivation; H is the village Hasanur.

42   *Main study area and methods*

Biligirirangan hills. The hills are separated from this perennial river by a cultivated tract. Among the important streams are the Nirdurgi and the Araikadavu. At their junction the waters are impounded by the Suvarnavati Reservoir (4 km² submersion area), an irrigation project. To the south a few streams drain into the perennial Moyar river, although their contributions are insignificant. The bulk of the waters of the Moyar originates in the Nilgiris. The Moyar discharges into the Lower Bhavani Reservoir (80 km² submersion area), an irrigation and hydro-electric project.

3.1.2   *Climate*

The study area shows a striking diversity of climate due to the varied relief and topography. Absolute maximum and minimum temperatures range from 40 °C in the Moyar valley (250 m) during April to below 0 °C in the open grasslands (>1600 m) during December–January. Average annual rainfall similarly varies from 50 cm in the Moyar valley to 185 cm on the mountain summit. Monthly rainfall at Hasanur (900 m) is shown in Fig. 3.2. Rainfall was normal over the area in 1981. In 1982 the area received only about half the normal rainfall.

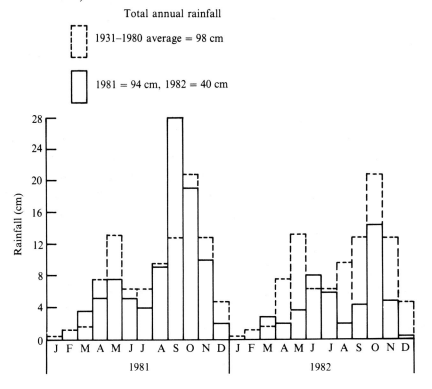

Fig. 3.2. Monthly rainfall at Hasanur (900 m elevation). (From Sukumar 1989a.)

The study area

More than any other climatic factor, it is the availability of water (both its direct availability for consumption and its impact on the vegetation) that determines the elephants' seasonal cycles. Three seasons may be defined, based mainly on the rainfall pattern and water availability in the environment.

(a) Dry season (January–April)
This period is characterized by negligible rainfall during January and February. There is some rain in March and April but this is still a period of 'hydrological drought', i.e. low stream flow and pond levels as defined by von Lengerke (1977).

(b) First rainy or wet season (May–August)
The heavy pre-monsoon showers in May and the subsequent influence of the south-west monsoon until August bring about a quantitative change in water availability. Rainfall attains a peak in August at the higher altitudes to the north of the study area.

(c) Second rainy or wet season (September–December)
Before the end of the southwest monsoon there is a sharp increase in rainfall over the southern part of the area, with a peak in October or sometimes September. Although rainfall from the northeast monsoon is irregular during November and December, sufficient water is available in the streams and ponds.

3.1.3   Land-use pattern

The area has been inhabited for many centuries by the Sholaga tribes who earlier practised shifting cultivation on the hills. Later, portions of the fertile valleys and plateaux were brought under permanent cultivation, chiefly of millets which depend largely on the rains. The British opened coffee plantations on the higher hill slopes (above 1250 m) beginning in 1888. In recent years the Forest Department has raised plantations of eucalypts (*Eucalyptus* spp.) and, to a lesser extent, teak (*Tectona grandis*) and silver oak (*Grevillea robusta*). The area has been a source of numerous products including timber, fuel wood, pulp wood and a variety of minor products such as fruits, bark and honey. Sandalwood (*Santalum album*) is an especially valuable commodity. Domestic livestock (density of $28/km^2$) which graze in the forest include cattle, buffalo and sheep. Cattle camps holding a few thousand animals each are also maintained inside the forest.

3.1.4 Vegetation types

The basic tropical vegetation types as defined by Puri (1960) include montane stunted evergreen *shola* forest and grassland (78 km²), moist deciduous forest (157 km²), dry deciduous forest (756 km²) and dry thorn forest (106 km²). In addition, many plantation forests are scattered over different habitats.

Human impact has converted most of the primary forest into secondary forms. For instance, because of opening up of the canopy the moist deciduous forest now exists only as an intermediate between the primary moist and the primary dry deciduous types. The dry deciduous forests have been degraded into scrub in the vicinity of settlements. The study area was divided into 19 zones based on vegetation type, secondary stages, species composition, topography, altitude, location and other natural features such as streams (Fig. 3.3 & Table 3.1).

Table 3.1. *Classification of habitat zones in the study area*

Zone number	Area (km²)	Vegetation type	Trees/Shrubs	Undergrowth grasses	Altitude (metres)	Topography
1	55	Dry deciduous	*Anogeissus latifolia, Acacia sundra, Lantana camara*	Short grasses	700–800	Flat
2	32	Dry deciduous	*Anogeissus, Terminalia* spp.	Tall grasses, *Themeda triandra*	800–1000	Sloping westwards
3	125	Mixed deciduous	*Terminalia tomentosa, Kydia calycina, Anogeissus*	Tall grasses, *Themeda cymbaria*	1000–1400	Hilly
4	78	Evergreen *shola* forest and grassland	*Elaeocarpus, Meliosma microcarpa*	Tall grasses, *Cymbopogon* in the grassland	1400–1800	Hilly, steep
5	50	Dry deciduous	*Anogeissus*	Tall grasses, *Phoenix humilis*	1000–1400	Sloping eastwards
6	37	Dry deciduous	*Anogeissus, Phyllanthus emblica*	*Dendrocalamus strictus*, tall grasses, *Themeda triandra*	900–1000	Sloping, southwards and westwards aspects
7	60	Riparian fringing and deciduous	*Anogeissus, Ziziphus* spp.	Short grasses, *Bambusa arundinacea* along stream	800–950	Flat, small hillocks

Table 3.1—continued

Zone number	Area (km^2)	Vegetation type	Trees/Shrubs	Undergrowth grasses	Altitude (metres)	Topography
8	76	Dry deciduous	*Acacia* spp., *Ziziphus* spp.	Short grasses	900–1000	Undulating
9	21	Eucalyptus plantation	*Eucalyptus*	Short grasses	850–900	Flat
10	53	Dry deciduous woodland–scrub	*Anogeissus*, *Acacia* spp.	Short grasses	850–950	Flat
11	158	Dry deciduous, northern part (11A, area 62 km^2) more dry	*Anogeissus*, *Pterocarpus marsupium*	*Phoenix humilis*, *Dendrocalamus*, tall grasses, *Themeda* and *Cymbopogon*	950–1300	Hilly
12	27	Riparian fringing and deciduous	*Acacia* spp. *Lantana*	*Bambusa* along stream, short grasses and sedges	900–1000	Valley
13	28	Dry deciduous	*Anogeissus*	Tall grasses, *Themeda* spp.	1000–1400	Hilly
14	32	Mixed deciduous	*Terminalia tomentosa*, *Kydia calycina*	Tall grasses, *Themeda cymbaria*	1200–1350	Undulating valley
15	44	Dry deciduous	*Terminalia*, *Kydia*, *Anogeissus*, *Phyllanthus*	Tall grasses, *Themeda* and *Cymbopogon*	1100–1300	Undulating
16	12	Grassland and Eucalyptus plantation	*Eucalyptus*	Planted *Bambusa*, tall grasses, *Cymbopogon*	900–1100	Undulating
17	136	Dry deciduous	*Anogeissus*, *Albizia* spp.	Tall and short grasses, *Themeda* spp.	400–1100	Steep slopes, southwards aspect
18	44	Dry thorn	*Albizia amara*, *Acacia latronum*	Short grasses	250–400	Flat
19	62	Dry thorn and riparian	*Albizia amara*, *Acacia* spp. *Hardwickia binata*, *Gyrocarpus jacquini*	Short grasses	250–400	Flat, valley

From Sukumar (1989a).

46 *Main study area and methods*

Fig. 3.3. Map showing major vegetation types and habitat zones in the main study area. (Modified from Sukumar 1989a.)

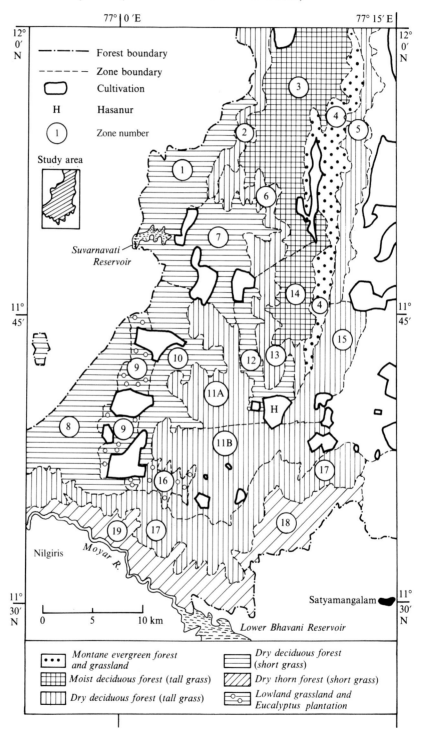

The study area

For the purpose of studying the elephant's pattern of habitat utilization in relation to feeding it was useful to classify the zones under these broad habitat types (Figs 3.4–3.8).

(a) Short grass areas with predominantly browse vegetation (Zones 1, 7, 8, 9, 10, 12, 18 and 19, total area = 398 km^2, entire area accessible).

(b) Mixed tall grass and browse in deciduous forest (Zones 2, 3, 5, 6, 11, 13, 14, 15 and 17, total area = 642 km^2, accessible area = 475 km^2).

(c) Lowland grassland (only Zone 16 which has been partly modified by plantations but taken as representative, area = 12 km^2, entire area accessible).

(d) Montane stunted evergreen *shola* forest and grassland (Zone 4, total area = 78 km^2, accessible area = 53 km^2).

The total area of a particular habitat is not necessarily accessible to elephants. For instance, a large proportion of the steep slopes of Zones 2 and 17 is not readily negotiable. Therefore, the accessible area is also given for making an unbiased comparison.

Fig. 3.4. Dry thorn forest with short grasses in the Moyar valley (Zone 19). In the background are the steep slopes (Zone 17) of the Eastern Ghats.

Fig. 3.5. Dry deciduous forest with undergrowth of *Phoenix humilis* and tall grass *Themeda triandra* (Zone 11).

Fig. 3.6. Moist deciduous forest with tall grass *Themeda cymbaria* (Zone 3).

The study area 49

Fig. 3.7. Lowland *Cymbopogon* grassland planted with *Eucalyptus* trees (Zone 16).

Fig. 3.8. Montane stunted evergreen *shola* forest and grassland at the summit (Zone 4) with coffee plantations in the valley.

3.1.5 Mammals

The Ethiopian, Palaearctic and Indo-Malayan faunas of the Indian sub-continent are well represented in the study area. Only the more conspicuous large mammals will be mentioned here.

The elephant (*Elephas maximus*) is the dominant wild mammal in terms of biomass. It is distributed throughout the study area except on very steep hill slopes. The density of elephants has been estimated to average just over 0.5 elephant/km^2 with a seasonal variation in numbers from 365 to 700 within a 928 km^2 area. The social structure of the Asian elephant was observed to be similar to that of the African elephant (Buss 1961; Laws, Parker & Johnstone 1975; Douglas-Hamilton 1972; Douglas-Hamilton & Douglas-Hamilton 1975; Moss 1982). The basic unit of elephant society is the 'family' consisting of an adult cow and her offspring, including daughters of all ages and sons of pre-pubertal age (Fig. 3.9). A joint family is composed of two or more adult cows, which are presumably sisters or mother and daughters, and their offspring. A family or joint family forms temporary bonds with one or more families to which they may be related. This level of organization has been

Fig. 3.9. A typical elephant family unit. The matriarch is at the extreme left.

The study area

termed the 'kin group'. During certain seasons a large number of elephant families group together temporarily. These 'clans' were obvious during the dry season in the study area. Male elephants disperse from the family around the pubertal age and move solitarily or form temporary bull groups (Fig. 3.10). Adult bulls associate with the family groups temporarily for mating with cows in oestrus. The terms 'family group' and 'family herd' used in this book include all levels of female-led social groups. The mean group size of family herds in the study area was similar during the dry season (8.2) and second wet season (8.8) but significantly different during the first wet season (5.8). Adult bulls were mostly solitary (93% of sightings). The largest bull group seen had three animals. Bulls associated with family groups on an average for 23% of their time.

Gaur (*Bos gaurus*) are seen mostly in the moist deciduous forests of Zones 3 and 14 (Fig. 3.11). About 100 to 200 gaur (average density $0.16/km^2$) may utilize the main study area. Spotted deer or chital (*Axis axis*) occur in sizeable numbers in the thorn jungle of the Moyar river valley. Elsewhere only small herds are found, usually near human settlements. The total population size

Fig. 3.10. A male elephant, in the 15–20 year age class, which has dispersed from its family.

was estimated to be 600–900 animals. Sambar (*Cervus unicolor*) are widely distributed over all habitat types, although only solitary animals or small groups are seen. A guess estimate of 500–1000 sambar has been made. Muntjac or barking deer (*Muntiacus muntjac*) are seen usually solitarily or in pairs mainly in moist deciduous forest. Between 300 and 600 muntjac may be found in the area. Blackbuck (*Antilope cervicapra*) are confined to the Moyar valley, especially the open area adjoining the Lower Bhavani Reservoir. A partial count gave an estimate of 400–600 blackbuck for the study area.

The major carnivores include the tiger (*Panthera tigris*), the leopard (*Panthera pardus*), the wild dog (*Cuon alpinus*) and the jackal (*Canis aureus*). The striped hyaena (*Hyaena hyaena*) is confined to the Moyar valley and the dry tract along the west (Zone 8). Other mammals in the area include the wild pig (*Sus scrofa*), the sloth bear (*Melursus ursinus*), the bonnet monkey (*Macaca radiata*) and the common langur (*Presbytis entellus*). A small herd of feral buffaloes (*Bubalus bubalis*) is found in the Moyar valley.

Fig. 3.11. A female gaur, *Bos gaurus*, the second largest mammal in the study area.

The study area 53

3.1.6   *The people*

About 16 000 people live in the settlements within the study area. In addition, at least 50 000 people living along the periphery also make an impact on the forest. Only the major human communities are listed below.

The Sholagas are a relatively primitive tribe who have lived in the area for many centuries (Fig. 3.12). There are many sects of the Sholagas. The forest dwellers are primarily food gatherers subsisting on honey, tubers, fruits and green shoots of plants. They also consume the meat of certain animals, such as deer and hare, from trapped or dead animals. They have practised shifting cultivation in the past but in recent times have been brought together under permanent settlement schemes. Currently, the Sholagas cultivate government land or their own small land holdings, work as labourers for wealthier communities and are engaged in collecting minor forest produce. The Irulas are also a forest tribe immigrant from the Nilgiris in small numbers to the Eastern Ghats. The Kadu Kurubas (*kadu* = forest) are similar to the

Fig. 3.12. A Sholaga family in front of their dwelling. The Sholagas are the original inhabitants of the area.

Sholagas in their life style. One sect, known as the Betta Kurubas (*betta* = hill), has taken to agriculture and wage earning, while another, the Jenu Kurubas (*jenu* = honey) are basically food gatherers. The Lambadis are a gypsy tribe immigrant from northern India some centuries ago. Lambadi women are unmistakable in their gaudy, fancy dresses and heavy ornaments of metal and ivory on their arms and legs, although such styles have waned considerably in recent years. Their main occupation is agriculture.

The Lingayats are a large, sectarian caste originally from the Karnataka plateau to the north. They shun meat and alcohol. Their main occupation is agriculture; they are also professional cattle breeders. The Lingayats are the most numerous among the communities in the study area. The Badagas in the area came from the Nilgiris, probably during the nineteenth century. British coffee planters have sold their estates to Indians. Government agencies such as the Forest Department have posted their staff in some villages.

3.2 Study methods
3.2.1 *Registration of elephants*

A photographic file was maintained on elephant family groups and adult bulls. Elephants were identified by characteristics of their ears such as cuts, holes and degree of folding, and in males the characteristics of tusks such as size, shape and broken ends (Douglas-Hamilton & Douglas-Hamilton 1975). Instances of sightings of identified elephants were marked on a map for estimating home range sizes. Peripheral locations of sightings were connected by straight lines to obtain the smallest convex polygon which contained all locations and the enclosed area was taken to be the home range (Jennrich & Turner 1969; Leuthold 1977; Olivier 1978*b*).

3.2.2 *Density estimates in different zones*

The habitat zones were taken as the units of sampling. Mean elephant densities in different zones during two-month periods were determined by ground transects, details of which are given in Appendix I. The purpose was to get a picture of the spatial distribution patterns of elephants during different seasons, from which inferences could be made on broad movement patterns and strategy of habitat utilization.

3.2.3 *Ageing techniques*

A large data base on captive elephants kept under semi-natural conditions in southern India was used to determine various parameters of growth. Growth equations based on von Bertalanffy functions were derived for shoulder height, body weight and circumference of tusk at lip line with age (Hanks 1972b; Laws *et al.* 1975). The relationships between height and weight, height and circumference of front foot, and tusk weight and tusk circumference were also determined (Appendix II).

The shoulder heights of wild elephants were estimated by a photographic method (Appendix II). Male elephants up to 20 years old and female elephants up to 15 years old were placed in age classes by relating their heights to the growth curve. Older elephants were classified by comparing their morphological features, such as degree of ear folding and buccal/temporal depression, with those of domestic elephants of known age. The age structure of the elephant population was constructed, incorporating family herds and adult bulls in the proportion in which they were present in the population.

Family groups were sampled in 1981, 1982 and 1983, avoiding repetitive sampling during a year. The members of each family group were aged at the first instance that they were clearly photographed. During 1982 and 1983, a family group was included only if it was sampled at least eight months after the previously sampled date. For adult bulls only an average structure for 1981–83 was taken, based on a sample of 24 identified elephants.

3.2.4 *Observations on feeding*

Direct observations of feeding by elephants were made in selected zones representing each of the three broad habitat categories. The scan sampling method was employed; the plant species being eaten by each visible member in a herd were recorded at 5-minute intervals (Altmann 1974). Each record was scored as browsing or grazing and the species noted. In some zones where sufficient direct observations were not made, an indirect method was used to determine the proportions of different browse plants in the diet. Ten plots, each measuring 0.5–2 ha depending on plant density in the zone, were examined for signs of elephant feeding on trees and shrubs. The number of individuals of each plant species and the number of branches broken from each individual were recorded.

Calculations were made as follows. Data on feeding were lumped together for all zones categorized under the three broad habitat types for the three major seasons (dry, first wet and second wet) and under each category the proportions of browsing and grazing determined. The seasonal densities of elephants in these habitats were used to give appropriate weighting to the diet of the elephant population in the study area as follows.

$$F_w = \sum e_i f_i,$$

where F_w is the weighted proportion of browsing or grazing during the season by the elephant population in the study area; e_i is the proportion of elephants of study area within the ith habitat type during the season; and f_i is the proportion of browsing or grazing within the ith habitat during the season.

To determine the total quantity of food consumed daily, the mean weight of one mouthful of grass was estimated and the rate of feeding and the total time spent in feeding daily were obtained from the literature. When a grass tussock is mature, elephants consume only the basal portion and discard the upper leaf blades. A random sample of 300 clumps of the discarded top leaves was collected and the mean number of leaves per clump consumed (c) calculated. Samples of 40 intact clumps were collected during the second wet season and the dry season; the entire clump was weighed and the basal portion cut similarly to the elephant's manner of feeding and weighed. This gave the mean weight of the basal portion of one leaf (b). Then the mean weight of grass consumed per mouthful is $b \times c$. All weights were taken fresh in the field and samples retained for determining oven dry weight. Unless otherwise stated only dry weights are reported in this book.

3.2.5 Primary production of grass

The net aerial primary production of grasses was estimated during 1982–83 by the harvest method (Singh, Lauenroth & Steinhorst 1975). Two sites of 50 m × 50 m each were located, one in the deciduous forests of Zone 3 (high rainfall) and the other in the grassland of Zone 16 (medium rainfall). At intervals of two months all the grasses, both live and current dead, in 30 random sub-plots of 1 m² each were clipped at ground level. The clumps were weighed fresh in the field and samples oven dried. The positive increments in the total biomass, representing the net primary production, were related to the total rainfall during the same interval (Whittaker 1970; Phillipson 1975; Sinclair 1975).

3.2.6 Herbivore biomass and grass consumption

Estimates of population sizes for the major mammalian herbivores were based on transect counts (elephant, gaur), on total counts in some locations (blackbuck, spotted deer), Forest Department records (domestic livestock) or guess estimates (other mammals). The population mean body weight is based on Schaller (1967) except for elephant, which was calculated from the age structure and age–weight regression equations (Appendix II). Daily food requirement was taken to be 1.5% (dry weight) fodder of the live

Study methods 57

body weight for elephants and 2.5% for other herbivores (Laws *et al.* 1975; Sinclair 1975). For elephants the calculations are based on densities in each zone at two-month intervals and the proportion of grass in the diet suitably weighted according to seasonal differences in the zones. Sambar was taken to be 20% grazer and all other herbivores as 80% grazers on average. Consumption by livestock pertains to only 10 months of grazing in the forest annually; they are usually left for grazing in the harvested fields for two months during the dry season.

3.2.7 Damage to trees

The impact of elephants on some commonly browsed trees was evaluated. Ten plots of 400 m^2 each were examined in Zones 3 and 15 for damage to *Kydia calycina* and *Grewia tiliaefolia*, and another ten plots of 0.5 ha each in Zone 12 for damage to *Acacia leucophloea* and *A. suma*. The stem girth, height and canopy volume were measured. The extent of damage to bark, banches and main stem was assessed, and the tree classified as living or dead. The annual damage to *Eucalyptus* plantations was assessed by numbering 602 trees in two plots of 0.25 ha each in Zone 16 during April 1982 and re-censusing the plots in April 1983.

3.2.8 Damage to crops

To study the pattern of crop depredation by elephants, 22 settlements, comprising 10 enclaves of cultivation (total area 46 km^2) of various sizes within the study area, and 2 settlements just outside the forest boundary, were monitored at least once a week from March 1981 to February 1982 and less intensively for another year for a qualitative comparison. Direct observations were also made on raiding elephants at night. Records of the name of the cultivator, date of raiding, the crops damaged and the number of elephants (either adult bulls or family herds) involved were maintained by local contact persons and collected when the villages were visited. These were used in computing the monthly frequency of raiding and the mean raiding sizes separately for bulls and herds. The mean raiding size was defined as the average of the total numbers of elephants visiting a village on sampled dates, only positive instances being considered.

If crop raiding was reported, an attempt was made to trace the path of the elephant group within the cultivated tract during a particular night and all the affected fields were inspected. In millet and cereal fields the quantity of crops consumed by elephants was estimated in the following way. Mean plant densities in the undamaged and the damaged portion of the field were obtained by laying 5–40 quadrats (depending on size of damaged patch) of

50 cm × 25 cm in each portion. The difference gave the mean number of plants (n) consumed per unit quadrat area. Since elephants consumed only the top portion of most cultivated grasses, an estimate of the mean weight of a plant consumed (w) was obtained from the difference in weights of random samples of 30 intact plants and 30 discarded basal parts. Thus, the quantity of crops consumed (nw) per unit quadrat area was extrapolated for the entire damaged area of the field. The quantities consumed from each field were summed to arrive at the total quantity eaten during one night of raiding by the bull group or family herd.

Although a total record was obtained from March to August, it was possible to inspect damage to fields for only a sample of raids from September to January when raiding was very frequent in most villages for the finger millet crop. In all, estimates of quantity consumed from fields were made for 77 nights of raiding by adult bulls involving 131 elephants and for 33 nights of raiding by family herds involving 297 elephants. From these estimates of mean quantity consumed per night of raiding by an elephant and the records of monthly raiding frequency and mean raiding sizes of bulls and family herds in different villages, the total quantity of crops consumed by the elephant population during each month was calculated separately for bulls and herds. All results are expressed in dry weight.

The economic loss was computed by taking the 1982 market value of the potential crop yields that were depressed by elephants and adding a standard 15% of the value as costs of cultivation. For coconut the loss took into account the period needed for replacement of destroyed trees. A certain number of 'man hours' are lost because farmers are forced to guard their fields at night. The wage potential that could be earned by each farmer during one hour each night for 100 nights in a year was computed.

3.2.9   *Chemical analysis*

Important wild and cultivated food plants were analysed for crude protein by the standard Kjeldahl method and ashed samples for calcium, sodium, magnesium and iron by an atomic absorption spectrometer. With the available time and resources the analysis was restricted to only those plant parts likely to influence diet selection for particular nutrients.

3.2.10   *Collection of records*

Details of manslaughter by elephants were obtained by questioning villagers during habitat surveys in southern India. In addition, instances of manslaughter during the study period were examined in more detail.

Study methods

Every elephant found dead is required by regulations to be examined by a veterinarian and the cause of death ascertained. About 600 post-mortem reports pertaining to elephant deaths during 1977–87 in the states of Karnataka and Tamilnadu were scrutinized. Twenty carcasses were also examined during the study. The age of these elephants was estimated from measurements of height, circumference of front foot and tusks. Statistics on forest land use were obtained from the Forest Department. A socio-economic survey covering 700 families was carried out in the study villages to collect data on agricultural land use.

4

Movement and habitat utilization

Animal movement can be considered from diurnal to life-time scale. A large mammal like the elephant can be expected to move considerable distances even within a short period. For the purpose of management it is important to consider not the daily movement but the seasonal strategy of habitat utilization. A knowledge of the home range area of a species is also necessary in planning for nature reserves of adequate size.

4.1   Elephant densities and seasonal distribution patterns

Data on elephant densities in different zones covering an area of 928 km^2 are presented in Appendix I. Elephant densities are depicted in Fig. 4.1 to show the spatial distribution patterns during January–February, March–April, July–August and November–December. Except for January–February, the data have been combined for 1981 and 1982 since there was not much difference in the zonal density patterns during corresponding periods in these two years.

(a) Dry season distribution (January–April)
Certain riverine vegetation zones had the highest elephant concentrations during the dry season. Zone 12 had a density of about 4 elephants/km^2 during 1981 and 1982 when some water was available in the Araikadavu stream. In the Moyar river valley (Zone 19) there was a high concentration during the early part of the dry season, but elephants began dispersing from here in March. The thorn forest in the plains (Zone 18) was also utilized intensively during January–February but not afterwards.

Considerable numbers of elephants were also present in some of the tall grass forests. The grassland Zone 16 and the deciduous forests of Zone 11B had high to medium densities during January–February; Zones 3 and 6 had

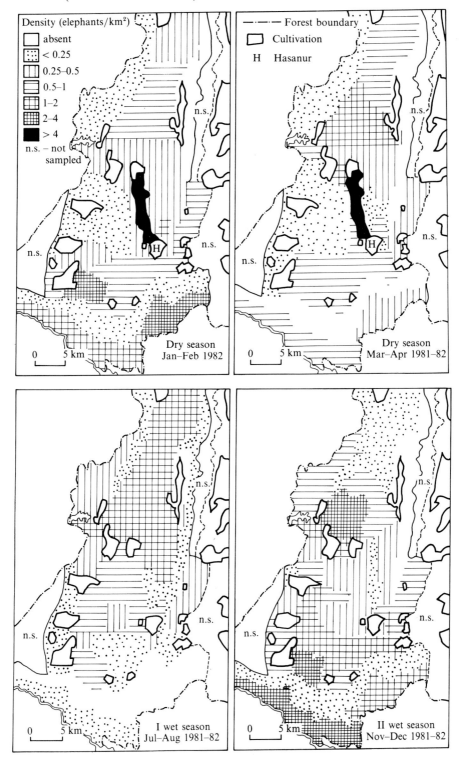

Fig. 4.1. Seasonal distribution patterns of elephants in the study area. (From Sukumar 1989a.)

medium densities in March–April. All other zones showed a low density or absence of elephants during this season.

(b) First wet season distribution (May–August)
After the pre-monsoon showers in April–May, elephants dispersed from the riverine habitats occupied during the dry months. For instance, in Zone 12 there was a sharp fall in density to 0.5–1.0 elephant/km^2, a mere fraction of the dry season concentration. There were few elephants in the Moyar river valley (Zone 19) during May–August. But elephants utilized one riverine habitat (Zone 7) at a higher intensity than earlier.

The zones of tall grass forest experienced an influx of elephants. This was most noticeable in the deciduous forests of Zones 3 and 6, where a density of 1–2 elephants/km^2 was reached. Because elephants were now diffused over a wider area, no zone had a very high density such as occurred during the dry season. Between May and August the total number of elephants in the main study area was reduced owing to movement to outside areas. Such a seasonal excursion was mainly from the Moyar valley southwards and westwards into the Nilgiri hills.

(c) Second wet season distribution (September–December)
Beginning in September there was a distinct movement from Zone 3 southwards into the lower-elevation short grass habitat of Zone 7. Elephants also increasingly occupied the scrub woodlands of Zones 8 and 10. In late October there began an influx into the Moyar valley from the Nilgiris and probably also from the Talamalai plateau through a few passes in the steep hills.

In September–October the situation was still somewhat confused, but by November–December certain well-defined concentrations could be seen. These included a high density of 2–4 elephants/km^2 in Zones 7 and 19, and about 2 elephants/km^2 in Zone 16 and adjacent areas of Zones 8 and 11B. Elephants also moved into Zone 18, beginning in November.

4.2 Movement pattern of different clans

Based on the dry season distribution, certain distinct clusters of numerous elephant families could be identified in different regions of the study area. These aggregations were similar to the 'clans' described for the African elephant (Douglas-Hamilton 1972; Laws et al. 1975; Moss 1982). These clans may consist of many related elephant families, as presumed in the African species, although no firm evidence for this can be presented here. At least five such clans with overlapping home ranges were present in the study

Home range sizes 63

area. Each clan or aggregation seemed to consist of about 50–200 individuals as inferred from elephant density estimates in zones where such aggregations occurred.

Families of a clan seemed broadly coordinated in their seasonal movements. This was inferred from resightings of identified families. For instance, certain families of Clans 1 and 2 occupied the same zones during the corresponding season of different years. Clans also showed distinct seasonal movements in the study area. Similar movements were observed by McKay (1973) in some elephant populations in the dry southeastern region of Sri Lanka. Khan (1967), who followed the movements of a single herd in Malaysia, found no seasonal pattern. This could be due to the relatively aseasonal climate in equatorial rain forest and a certain constancy in the availability of food and water. Leuthold (1977) observed that individual elephants or elephant groups in Tsavo, Kenya, made occasional sudden movements over distances of 30–50 km within a few days. Without radio-tracking such movements could not be detected in the study area.

4.3 Home range sizes and factors influencing them

The home range sizes of family herds of Clan 1 and Clan 2 were about 105 km^2 and 115 km^2 respectively (Table 4.1 and Fig. 4.2). Since the other clans moved out of the study area their home range sizes could not be

Table 4.1. *Resightings and home range sizes of elephants*

Elephants	Number of identified sightings	Zones in which recorded	Time (months)		Linear distance of farthest resighting (km)	Home range size (km^2)
			Between farthest resighting	Between first and last sighting		
Adult bull (MA-6)	11	9, 11, 12, 16 and in cultivation	7.3	19.5	21	170
Adult bull (MA-19)	12	19, also in Mudumalai and Bandipur to west of main study area	5.2	26.0	52	320
Adult bull (MA-25)	7	8, 9, 10 and in cultivation	3.0	9.3	20	215
Family herd (Clan 1)	14	3, 6, 7	16.4	23.5	25	105
Family herd (Clan 2)	15	7, 11, 12, 16	18.0*	23.2	21*	115

* A movement of 17 km within 4 months was also recorded.
From Sukumar (1989a).

determined. For three identified adult bulls, the home range sizes were 170 km² (MA-6), 320 km² (MA-19) and 215 km² (MA-25). The farthest linear distance of resighting was 21 km for a family herd and 52 km for a bull.

The elephants in the study area are part of a larger population distributed over a contiguous area of 10 000 km². The estimated home ranges of about 100–300 km² should be regarded as minimum sizes. Leuthold (1977) found that home range sizes determined by radio-tracking were usually larger than those revealed by visual identification.

Although it can be intuitively expected that an animal's range of movement will increase with greater body size and energy requirement (McNab 1963), the predictive equations in the literature (Peters 1983; Mace, Harvey & Clutton-Brock 1983) give no indication of the enormous variation in home range size of a species such as the elephant. The following broad patterns can be seen with regard to factors influencing the range sizes of different elephant populations.

Home range sizes of only 14–52 km² for the African elephant in Lake Manyara National Park, Tanzania (Douglas-Hamilton 1972), seem to be imposed by barriers to free movement. The unpredictable availability of

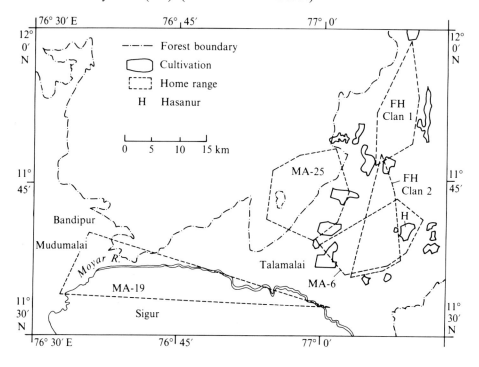

Fig. 4.2. Home ranges of some identified adult male elephants (MA) and family herds (FH). (From Sukumar 1989a.)

water in semi-arid zones may contribute to relatively large home ranges. In the dry Tsavo region of Kenya, range sizes of over 2000 km² have been reported (Leuthold 1977). In areas where water is not a limiting resource, the availability of food may govern the home range size. In the rain forests of Malaysia, Olivier (1978b) found that the home range size of family groups was larger in primary forest (up to 167 km²) where food plants are less abundant than in secondary forest (up to 59 km²). The diversity in habitat types may also influence range size. The more diverse a region, the smaller could be the home range since elephants would be able to meet their varied seasonal requirements within a relatively restricted area.

4.4   Inter-annual differences in movement pattern

When environmental factors, especially rainfall, show any drastic changes, the normal movement pattern may be upset. Evidence for this came from the dry-season elephant distribution of 1983. The patterns described so far were largely true for 1981 and 1982. In 1982 the annual rainfall was only about half the normal amount. Environmental effects were not evident immediately, but later the dry season of 1983 was an especially harsh period.

During the normal years a very high elephant concentration was seen in Zone 12 between January and April (Appendix II). In 1983 there was practically no water in the Araikadavu stream during these months. During January–February 1983 the density of 1.4 elephant/km² was much lower than the 1982 density of 5.0/km². By March–April 1983 there were hardly any elephants in the area, as the following comparison reveals.

Density of elephants/km² in Zone 12

Year	January–February	March–April
1981	not sampled	4.3
1982	5.0	3.7
1983	1.4	0.3

Correspondingly, during March–April there was an increase in elephant concentration in the northern part of Zone 3, where a perennial pond became an important water source. Because these changes in distribution occurred at the end of the study, the specific shift in habitat of the clans involved could not be determined.

4.5   A long-term perspective on elephant movement patterns

Sanderson (1878), who captured elephants in this area during the nineteenth century, has made observations on their movement in the

66 *Movement and habitat utilization*

Biligirirangans which can be compared with the pattern today. The following lines are taken from his book.

> 'In the dry months – that is, from January to April, when no rain falls the herds seek the neighbourhood of considerable streams and shady forests. About June, after the first showers, they emerge to roam and feed on the young grass. By July or August this grass in hill tracts becomes long and coarse . . . elephants then descend now and again to the lower jungles, where the grass is not so far advanced . . . the herds invariably left heavy jungle about October for more open and dry country. About December, when the jungles become dry, and fodder is scarce, all the herds leave the low country, and are seldom seen out of the hills or heavy forests until the next rains.'

The observations of Sanderson correspond exactly to the pattern observed for Clan 1, which ranges over Zones 3, 6, 7 and possibly adjoining areas. This was the region with which Sanderson was familiar. Obviously, the movement pattern of the elephants in this area has not basically changed for over a century. In this long-lived mammal, the adult females are repositories of traditional knowledge, including migration routes. This could contribute to a conservative home range and movement pattern.

4.6 Seasonal use of habitat types

The mean elephant density in the study area was $0.56/km^2$ in 1981 and $0.53/km^2$ in 1982. Thus, for a given zone a seasonal density higher than about $0.5/km^2$ indicates a greater than expected use and a lower density a lower than expected use (see Appendix I for density values). Habitat use can also be considered for the broader vegetation types. The proportions of elephants in these habitat types in relation to the area available during 1982 are shown in Fig. 4.3.

During the early part of the dry season (January–February) the short grass habitats were clearly used more than their proportional availability. This declined progressively with the onset of pre-monsoon showers (March–April). During the first rainy season (May–August) the deciduous forests with tall grasses were generally preferred. But after the second heavier rains (September–December) elephants moved once again largely into the short grass open habitats. High-altitude evergreen *shola* forest and grassland were largely shunned. They were utilized to a negligible extent throughout the year. However, in another elephant region, the Nilgiris, it was noticed that elephants utilized this vegetation type to a greater extent during the dry months. The lowland grassy areas were used in a high proportion relative to their availability. In the Mudumalai sanctuary, a preference for the swampy grasslands was noticed during the dry season.

4.7 Movement in relation to foraging and availability of water

The movements of elephants in the study area are largely in conformity with expectations of optimal foraging theory (Pyke 1983; Sinclair 1983). These were related to nutritive values of food plants and water availability in different habitats (see Chapter 5 and Appendix III for more details).

(a) During the dry season, browse plants have a higher crude protein content (6–18% dry weight) than grasses (1.5–2.5% in the basal portion consumed during this season). Hence, elephants could be expected to show a preference for the predominantly browse vegetation zones, and also for browse plants within the tall grass forests. As seen in Fig. 4.3 they clearly utilize the browse habitats in greater proportion than availability.

(b) After the onset of rains, the tall grasses become highly palatable (8–10% protein) for a few months, especially in the fire-affected

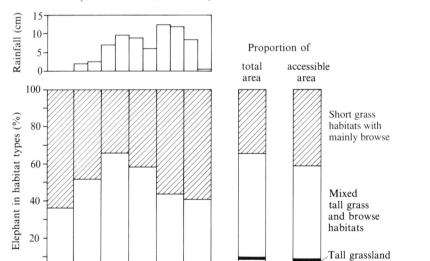

Fig. 4.3. Seasonal use of habitat types by elephants in relation to rainfall, 1982. (From Sukumar 1989a.)

areas. As expected, elephants show a distinct movement into the tall grass forests during the first wet season to feed mainly on grass.

(c) During the second wet season, when the tall grasses turn fibrous and siliceous with a lower palatability and protein content (2–4% in the basal portion), elephants again increasingly utilize the short grass habitats.

However, elephants need not just food but also water in large quantities. Their movement would also be governed by the spatial distribution and temporal availability of water. Studies on the seasonal distribution of the African elephant have shown high-density strata along water sources during the dry spell and a dispersal after the rains (Leuthold 1977; Allaway 1979). In the study population, one such aggregation occurred in a riverine habitat (Zone 12) during the dry months. But in another river valley (Zone 19), elephants, though present at a high density during the early dry period, began to disperse in March when water was even more difficult to obtain elsewhere. Obviously factors other than water influenced the movement of this clan. Elephants have to strike a compromise between food and water; both may not necessarily be available at their best in the same area.

5

Feeding and nutrition

It is known that elephants are generalist feeders, consuming a large number of plant species. Numerous studies on feeding habits of African and Asian elephants have shown that proportions of various plant categories in the diet vary widely from one region to another (see, for example, Wing & Buss 1970; Field 1971; McKay 1973; Laws *et al.* 1975; Field & Ross 1976; Guy 1976; Vancuylenberg 1977; Olivier 1978*b*; Barnes 1982; Sukumar 1985; Sukumar, Bhattacharya & Krishnamurthy 1987). A feeding pattern established for one area cannot be naively extrapolated to another area. A knowledge of feeding preferences and nutritive requirements is essential in planning for habitat management. This study on feeding in the wild in relation to seasonal chemical composition of food plants also provides the necessary background for understanding crop-raiding behaviour.

5.1 The plants and parts eaten

Within the study area 112 plant species eaten by elephants in the wild were recorded. Although the elephant is a generalist feeder, the most commonly eaten plants were from only five botanical taxa: the order Malvales (families Malvaceae, Sterculiaceae and Tiliaceae) and the families Leguminosae (especially the sub-family Mimosoideae), Palmae, Cyperaceae and Gramineae. These taxa accounted for 76 (68%) of the recorded food species. In fact, just 25 species from these taxa constituted about 85% of the elephants' quantitative intake. Other plant families that were important as food sources included Rhamnaceae, Anacardiaceae, Moraceae, Capparidaceae, Burseraceae, Rutaceae, Sapindaceae, Myrtaceae, Euphorbiaceae and Verbenaceae.

Elephants are both grazers and browsers. Grazing was primarily on the tall grasses *Themeda cymbaria*, *T. triandra* and *Cymbopogon flexuosus*. The portion of the grass consumed varies with season. After the first rains when the new flush appears, elephants remove the tender blades in small clumps without uprooting the plant. This was noticed during April–June. Later, when the grass attains a height of 0.5–1 m, entire clumps are uprooted with the trunk, dusted skilfully and the relatively fresh top portion of the leaves consumed, while the basal portion with the roots is discarded. When the grasses mature (October–March), the succulent basal portion with the roots is consumed after vigorous cleaning and the fibrous blades discarded. Short grasses were generally eaten only during the second wet season when they attained a height of about 0.5 m and began to flower (Fig. 5.1). Elephants may also selectively forage on the relatively tender short grasses growing beneath the coarse swards of tall grasses. In the dry zone of Sri Lanka, elephants remove grasses at a shorter stage by scraping the surface of the ground (McKay 1973).

From the bamboos *Bambusa arundinacea* and *Dendrocalamus strictus*, various portions such as seedlings, culms and lateral shoots are consumed (Fig. 5.2). When elephants are feeding on trees and shrubs, both leaves and

Fig. 5.1. Elephant bull feeding on grasses.

The plants and parts eaten

twigs are taken, as in species of *Acacia*, *Albizia*, *Grewia*, *Ziziphus* and *Ficus*. Elephants may select individuals within a species with the most fresh foliage, as was often observed when they fed on *Acacia pennata*. In the absence of leaves they still consume the twigs during the dry season. Thorn-bearing shoots of many species of *Acacia* are consumed without any obvious discomfort. Elephants feed on the bark of certain plants such as *Acacia suma*, *Grewia tiliaefolia*, *Kydia calycina*, *Helicteres isora*, *Ziziphus xylopyrus*, *Tectona grandis* and *Eucalyptus* spp. (Fig. 5.3). Among the fruits consumed are those of *Limonia acidissima*, *Tamarindus indica* and *Careya arborea*. Leaves and fruits of the shrubby palm *Phoenix humilis* are eaten. Young plants may be uprooted with the forefeet, dusted, the basal soft stem chewed and the leaves discarded. Succulents such as *Sansevieria* and *Pandanus* are favoured, although these were not abundant in the study area.

5.2 **Proportions of browse and grass in the diet**

The proportion of time spent in browsing and grazing seasonally in each of the three broad habitat categories is given in Table 5.1. Based on the

Fig. 5.2. Elephant bull feeding on bamboo, *Bambusa arundinacea*.

72 *Feeding and nutrition*

Fig. 5.3. Elephant bull feeding on bark of young *Tectona grandis* plant.

elephant occupancy in these habitat types, a weighted average proportion is also given.

(a) In the short grass zones, both during the dry season and first wet season, a very high proportion of the diet (85–90%) was browse; grazing was restricted to sedges and grasses growing along streams. Only after the short grasses had grown above 0.5 m during the second wet season were they consumed.

(b) In the mixed tall grass and browse forests there was only marginally more grazing than browsing during the dry months. But the new growth of grass after the first rains promoted significantly more grazing (73%) especially on *Themeda*. During the second wet season grazing on tall grass declined, although the data are insufficient to prove this.

(c) In lowland grasslands, grazing obviously remained at a high level (>80%) throughout the year, although in view of the very restricted area the overall contribution of grass from here was low.

Table 5.1. *Proportion of time spent in browsing and grazing*

Habitat type	Jan.–Apr. (Dry) Browse:Grass	May–Aug. (I Wet) Browse:Grass	Sept.–Dec. (II Wet) Browse:Grass
1. Short grass habitat with predominantly browse vegetation	90:10	87:13	71:29
2. Mixed tall grass and browse habitat	45:55	27:73	40:60
3. Predominantly grassland habitat	19:81	+	6:94
Weighted seasonal proportion in study area*	69:31	46:54	56:44

+ Relatively few elephants were observed. Intermediate value of 12:88 has been taken.
* The feeding proportions were weighted by the proportion of the elephant population of the study area within the habitat type in 1982 (Chapter 4, Appendix I).

Statistical tests:
(a) The proportions of browsing versus grazing in each habitat type during different seasons were tested for statistical significance by the z test. Those which were significantly different at least at the 10% level were the following.
  Habitat 1: dry/II wet ($p < 0.05$), I wet/II wet ($p < 0.10$).
  Habitat 2: dry/I wet ($p < 0.01$).
  Habitat 3: dry/II wet ($p < 0.01$).
(b) The proportions of browsing versus grazing in different habitats during each season were also tested for statistical significance. All the differences were significant ($p < 0.01$).
Modified from Sukumar (1989*b*).

Overall, the importance of browse (69%) during the dry season is clear. Feeding on grass (54%) picks up during the first wet season. Once again after the second heavy rains there is increased feeding on browse (56%) in the natural habitat. These seasonal differences in proportion of grazing and browsing are remarkably similar to the results of observations on feeding by African elephants in the Kidepo Valley National Park, Uganda (Field & Ross 1976).

Studies in Sri Lanka generally report that grass is predominant in the elephant's diet. For instance, elephants in the Ruhuna National Park spend about 86% of their feeding time on short grasses (McKay 1973). It is likely that some of these results are biased because observations were made in open grassy areas without considering occupancy of different habitats (cf. Vancuylenberg 1977). The time of observation may also be important. At Gal Oya elephants predominantly graze in the morning before 09.00 and in the afternoon after 15.00, presumably when they come out into open areas, and the proportions of grazing and browsing are roughly equal between these hours (McKay 1973). Grasses are highly preferred in the rain forests of Malaysia, although their contribution to the overall diet is much less than that of browse owing to their low availability (Olivier 1978*b*).

5.3 Proportions of different browse plants in the diet

Important browse plants consumed in 5 representative zones in the study area are given in Table 5.2. These include two tall grass habitats (Zones 3 and 11) and three short grass habitats (Zones 12, 18 and 19). The Malvales and the Leguminosae (particularly *Acacia*) predominate in the browse diet of elephants in the study area. It is well known that various species of *Acacia* are also important in the diet of elephants in East Africa (see for example, Field & Ross 1976).

Elephants in the dry eastern region of Sri Lanka seem to have a more varied browse diet. Mueller-Dombois (1972) enumerated the proportion of stems in the 2–5 m height class of species utilized by elephants in the Ruhuna National Park. Ishwaran (1983) calculated preference indices for different woody plants by relating proportions of stems damaged to the proportions of their availability in Gal Oya National Park and the Himidurawa Tank region. Table 5.3 lists 55 species from 23 families which are important in the elephant's diet in these regions.

Palms dominate the diet of elephants in the rain forests of Malaysia (Olivier 1978*b*). There are at least 345 species of wild palm, most of which would be potential food, in these forests. Other plant families in which Olivier (1978*b*) recorded numerous species of food plants are Euphorbiaceae

Table 5.2. *Proportions of important browse plants consumed in the study area*

Plant species	Zone 3	Zone 11	Zone 12	Zone 18	Zone 19
Malvales					
1. *Kydia calycina*	56.1	R	—	—	—
2. *Helicteres isora*	12.6	—	—	—	—
3. *Grewia tiliaefolia*	17.9	9.9	2.3	—	—
Leguminosae					
4. *Atylosia albicans*	—	5.3	—	—	—
5. *Tamarindus indica*	—	R	8.4	5.0	R
6. *Hardwickia binata*	—	—	—	—	7.5
7. *Dichrostachys cinerea*	—	—	R	9.0	R
8. *Mimosa rubicaulis*	—	3.0	R	—	—
9. *Acacia leucophloea*	—	R	3.4	14.4	15.0
10. *Acacia latronum*	—	—	—	13.1	R
11. *Acacia suma*	—	—	12.4	—	—
12. *Acacia sundra*	—	—	R	20.3	12.2
13. *Acacia ferruginea*	—	—	—	3.6	—
14. *Acacia torta*	R	14.4	5.6	—	—
15. *Acacia pennata*	—	10.6	26.4	—	—
16. *Albizia amara*	—	—	—	23.0	49.7
Other dicotyledons					
17. *Capparis sepiaria*	—	—	2.3	—	—
18. *Commiphora caudata*	—	—	—	R	2.0
19. *Ziziphus xylopyrus*	R	5.3	3.4	5.4	8.8
20. *Tectona grandis*	R	3.8	R	—	—
Palmae					
21. *Phoenix humilis*	—	11.4	—	—	—
Bamboos					
22. *Bambusa arundinacea*	3.0	6.8	14.6	—	—
23. *Dendrocalamus strictus*	R	10.6	R	—	—
Total percentage of above species	92%	82%	81%	96%	96%
Sample size (*n*)	246	132	178	222	147

Only those species which constitute at least 2% of the browse diet in a zone have been included.
R: Recorded as eaten but in negligible quantity (below 2%) in the zone.
—: Not found in the zone or not recorded being eaten.
From Sukumar (1989*b*).

Table 5.3. *Important woody plants consumed by elephants in Sri Lanka*

Family	Species
Annonaceae	*Polyalthia korinti*
	*Polyalthia sp.
Boraginaceae	Carmona microphylla
	*Cordia gharaf
	*Cordia monoica
Capparidaceae	*Capparis divaricata
	Crateva religiosa
Erythroxylaceae	Erythroxylum monogynum
Celastraceae	*Elaeodendron glaucum
	Pleurostylia opposita
Ebenaceae	*Maba buxifolia
Euphorbiaceae	Aporusa lindleyana
	Cleistanthus patulus
	Croton klotzschianus
	Croton lacciferus
	Croton sp.
	*Dimorphocalyx glabellus
	Drypetes sepiaria
	Fleuggea leucocarpa
	Mallotus rhamnifolia
	Phyllanthus polyphyllus
	Securinega leucopyrus
Leguminosae	*Bauhinia racemosa
	Cassia auriculata
	Cassia roxburghii
	*Dicrostachys cinerea
Linaceae	Hugonia mystax
Loganiaceae	*Strychnos nux-vomica
	*Strychnos potatorum
Melastomataceae	*Streblus asper
Myrtaceae	Eugenia bracteata
	Psidium guajava
	Syzygium cumini
Ochnaceae	Ochna lanceolata

Proportions of different browse plants in diet

Table 5.3—*continued*

Family	Species
Rhamnaceae	*Ziziphus mauritiana*
	Ziziphus oenoplia
Rubiaceae	**Canthium coromandelicum*
	*Randia dumetorum
	*Randia malabarica
Rutaceae	**Atalantia monophylla*
	*Limonia acidissima
	*Glycosmis pentaphylla
Salvadoraceae	**Salvadora persica*
Sapindaceae	*Allophylus cobbe*
	Dimocarpus longan
	Lepisanthes tetraphylla
	Sapindus emarginatus
	Schleichera oleosa
Sapotaceae	**Manilkara hexandra*
Sterculiaceae	**Helicteres isora*
	*Pterospermum canescens
Tiliaceae	**Pityranthe verrucosa*
Verbenaceae	*Premna latifolia*
	Vitex pinnata

A species has been listed only if over 20% of total stems of the species were utilized by elephant (in Mueller-Dombois 1972) or if the damaged stems of the species constituted over 1% of the total number of damaged stems of all plants or if the preference index for the species was greater than 2.0 (in Ishwaran 1983). An asterisk against a species indicates that over 40% of stems were utilized or damaged stems constituted over 5% of total stems of all plants damaged or the preference index was greater than 5.0. The preference index for a species is the percentage damage in the species divided by the percentage availability of the species. An index >1.0 indicates a positive preference and <1.0 a negative preference for the species.

Based on Mueller-Dombois (1972) and Ishwaran (1983).

78   *Feeding and nutrition*

(25 species), Leguminosae (17), Moraceae (17), Guttiferae (13), Anacardiaceae (11), Annonaceae (10), Sterculiaceae (8), Rubiaceae (7) and Lauraceae (7). Fruits may also be important in the diet of rain forest elephants. Short (1981) found traces of fruit in 93% of elephant dung piles examined and recorded 35 species of fruit in closed-canopy forests of Bia National Park, Ghana. It can be seen that certain plant families are potential food sources for elephants, irrespective of vegetation type or region.

5.4   Quantity of forage consumed

Various estimates of quantity of food consumed by elephants have been reported. These range from 135 to 300 kg fresh weight for adults (a comparative account is given by Guy 1975). Laws *et al.* (1975) estimated on the basis of stomach contents that elephants consume 1.5% (dry weight fodder) of their body weight every day.

The mean clump size of tall grasses consumed by elephants in the study area was 85 leaves (s.d. = 47.8, n = 300 clumps). Because grass clumps discarded by juvenile elephants may have been under-represented or missed, the results refer to feeding by elephants aged above five years. The mean dry weight of one clump or one mouthful of grass was 57.8 g (s.d. = 22.4, n = 40) during the dry season and 76.5 g (s.d. = 22.7, n = 40) during the wet season.

Observations on feeding indicated a mean rate of 1.7 mouthfuls of grass per minute during the active phase of feeding. The more detailed records of Vancuylenberg (1977) gave a mean rate of 0.8 mouthfuls per minute during

Table 5.4. *Feeding rate and quantity of grass consumed*

	Dry season	Wet season
Mean weight of one mouthful	57.8 g	76.5 g
Feeding rate in mouthful/minute[a]	0.8	0.8
Quantity of grass consumed		
(a) per hour	2.8 kg	3.7 kg
(b) in 12 hours	33.6 kg	44.4 kg
Forage consumed in 12 hours as percentage of body weight[b]	1.5%	1.9%

[a] Feeding rate based on Vancuylenberg (1977).
[b] The body weight of an average elephant above 5 years in the population was calculated from the age–sex structure (Chapter 11) and age–body weight relation (Appendix II) to be 2700 kg.

the entire period of feeding. The total time spent in feeding may be between 12 and 19 hours per day (Wyatt & Eltringham 1974; Guy 1975; Vancuylenberg 1977).

Elephants can thus consume between 1.5% (dry season) and 1.9% (wet season) of their body weight in just 12 hours of feeding (Table 5.4). These data indicate that elephants may consume higher quantities than those reported by other workers (cf. Laws *et al.* 1975). Observations on crop raiding elephants also support these results. Adult bulls were capable of consuming up to 12 kg of crop plants per hour or 1.5% of their body weight in just 6–7 hours of intensive feeding, although this may not be representative of feeding in the wild. Some of the variation in estimates of quantity made by different workers could be due to reporting of data on a fresh weight basis. The moisture content of plants varies considerably even on a given day. Grasses had moisture contents varying seasonally from 40 to 80% of their fresh weight. It is also possible that estimates of feeding on grass alone cannot be extrapolated to the entire diet. The quantities of browse consumed per unit time may be lower, although owing to practical difficulties these cannot be accurately estimated.

5.5 Drinking

Elephants also need large quantities of water. An elephant may drink over 100 litres of water at one time and up to 225 litres in a day (Sikes 1971). Although an elephant may drink every day when water is available, the African elephant has shown a capacity to go without water for extended periods. A herd of 34 elephants which accidentally entered an enclosure on the Galana ranch, Kenya, went without any 'free drinking water' for 14 days before some of them broke out of the enclosure. Two juveniles, aged 2–3 years, which persisted, died on the fifteenth and the seventeenth day (Heath & Field 1974). Elephants in the Namibian desert go without water for prolonged periods (Walker 1982). These observations contradict previous assumptions that elephants need water every day. Comparable data on Asian elephants are not available.

When surface water is not available in streams or ponds, elephants dig holes on dry stream beds using their forefeet and trunk, and drink the sub-soil water which seeps into the cavity (Fig. 5.4). This behaviour was often noticed in the study area. An elephant was even observed inserting its trunk into its mouth, withdrawing fluid and spraying it on its body. This act, also reported by other workers, may have some survival value (Krishnan 1972; Douglas-Hamilton & Douglas-Hamilton 1975).

Elephants are fond of alcoholic liquor. There is anecdotal evidence of

80   *Feeding and nutrition*

elephants raiding the liquor stores of illicit brewers in the forest. This attraction may even lead them to damage huts in which liquor is stored. An adult bull (MA-25) drank a barrel of palm toddy in one of the study villages and went on a rampage in the area. Alcohol has an intoxicating effect on them, similar to the effect of fermentation of certain fruits consumed (Sikes 1971).

5.6   Nutrition and foraging

The diet of a herbivore is influenced by the various anatomical and physiological characteristics of the animal, the structural and chemical constitution of the plants, and the community structure (Owen-Smith 1982). Models of optimal foraging start with the basic assumption that the fitness of an animal depends on deriving the maximum energy within the minimum possible time (reviewed by Schoener 1971; Pyke, Pulliam & Charnov 1977). Herbivores need not only energy but also other nutrients such as protein or minerals and, in addition, have to avoid toxins in plants. A 'linear programming' model, which permits a simultaneous treatment of all these

Fig. 5.4. Elephants drinking sub-soil water from holes they have dug on the dry stream bed.

factors, seems the most appropriate for large generalist herbivores (Westoby 1974; Belovsky 1984).

5.6.1 Anatomy and physiology

The high-crowned (hypsodont) molar teeth with their rasp-like (occlusal) surface are structured for grinding fibrous and siliceous material. Though a non-ruminant, the elephant retains the advantages of cellulose digestion in the large caecum and the colon, which are sites of fermentation by microbes (McBee 1971; Clemens & Maloiy 1982). The food quality and digestive efficiency may be lower in a non-ruminant compared with a ruminant, but these may be offset by a higher throughput rate of food material. This means that a non-ruminant can tolerate a diet of lower quality but has to increase the feeding rate or the proportion of time spent in feeding.

The quantity of food and also the requirements of specific nutrients may be proportional to body weight, $W^{0.75}$ (Peters 1983). Thus, a large herbivore has a greater absolute food requirement than a smaller herbivore, but a lower requirement per unit body weight. The daily food requirement of an elephant has been estimated at 1.5–2% (dry weight) of its body weight (Laws *et al.* 1975) (Table 5.4), compared to between 2 and 3% for smaller African ungulates (Sinclair 1975). Even this smaller proportion translates into a large absolute amount needed daily by an elephant whose body weight may go up to 5 tonnes in an adult bull. To achieve this enormous intake the elephant has to be a relatively unspecialized feeder.

The prehensile trunk with the 'finger' at the tip enables the elephant to delicately remove a tiny *Mimosa* inflorescence or bring down a stout branch from a height of five metres. For its sheer range this capability is unsurpassed by any other terrestrial herbivore. The number of plants in the diet of the elephant includes 59 species at Kidepo Valley (Field & Ross 1976), 133 at Sengwa (Guy 1976) and 134 at Lake Manyara (Douglas-Hamilton 1972) for the African elephant; 89 species at Gal Oya (McKay 1973), about 400 in Malaysia by indirect registration (Olivier 1978b) and 112 in the present study for the Asian elephant. However, it must also be noted that the bulk of its diet may consist of only a few plant taxa.

The level of protein in the diet needed to maintain body weight is also related as $W^{0.75}$, but it may not be possible to extrapolate this relation all the way up to an elephant (Owen-Smith 1982). An elephant requires daily 0.3 g of digestible protein per kilogram of its body weight (McCullagh 1969). If it consumed 15 g dry matter per kilogram of body weight and 40% of the crude protein is digestible (Benedict 1936), a minimum level of 5% crude protein in plant material is needed for maintenance. It may be advantageous for an animal to lower its protein intake at certain times. The loss of body weight during the dry season is often viewed with concern. One of the reasons for avoiding excessive fat deposits during summer may be to have more efficient thermoregulation. During the dry period excessive protein intake is also undesirable, because nitrogen excretion requires more water which may be in short supply. It can be expected that animals are adapted to the recurring dry season, although a drought may decrease their survival.

Only crude estimates are available for requirements of two minerals, calcium and sodium. In male elephants increment in tusk weight is nearly linear at the rate of 1.4 kg/year or 3.8 g/day between the ages of 10 and 30 years (Sukumar 1985) (Appendix II). Because tusks are 45% calcium, this means a calcium need of 1.7 g/day for tusk growth alone. By extrapolating from human requirements, McCullagh (1969) calculated that a 1000 kg male elephant needs 8–9 g of calcium per day, which Laws *et al.* (1975) considered to be an underestimate. A pregnant or lactating elephant may have a high calcium requirement. Considering that an adult woman (say 65 kg) needs daily 20 mg calcium per kilogram body weight (based on U.S. Food and Nutrition Board Standard), a 3000 kg cow elephant may require 60 g per day. Female elephants are either pregnant or lactating for most of their adult life span.

The sodium budget of an elephant is likely to be precarious. Benedict (1936) estimated a net sodium deficit of 5 g/day for a captive Asian elephant; Olivier (1978*b*) calculated a deficit of 11 g/day for the wild elephant in Malaysia. From the intake–excretion figures given by them it would seem that an adult elephant requires 75–100 g sodium daily.

5.6.2   Palatability and nutritive value of foods

The proximate factor that influences the decision to consume or reject a plant is the palatability of the item as conveyed to the herbivore through the senses of taste, smell, sight and touch. Selection of tall grasses by elephants is related to their phenophase and palatability. During the first wet season, the new flush of grass, low in fibre and silica, is preferred by elephants. Fresh grass may also be palatable for its higher soluble carbohydrate content (Field 1976). When the grasses mature the fibre and silica contents increase. An examination of grasses at different seasons reveals their increasingly abrasive nature as growth progresses (values are given by Field (1971, 1976)). Elephants now switch over to the basal portion of the grass tussock, which is succulent. They also increasingly feed on short grasses, which are relatively tender.

Ultimately, the diet should provide all the nutrient requirements of the animal. Ungulates show a positive selection of plant species and plant parts with the highest protein value (Field 1976) or minerals such as sodium (Belovsky 1981). The elephant's strategy of alternating seasonally between grass and browse is related to the plants' protein content (see Appendix III for values). Fresh growth of tall grass during the first wet season had a crude protein content of 8–10% dry weight. Leaves of *Themeda cymbaria* had 10% protein in May and this altered only slightly to 8% at the end of the second wet season. By the dry season it dropped considerably to 3% protein. *Themeda* bases had less protein, with 4% in the wet season and 1.6% in the dry period. *Cymbopogon flexuosus* also showed a similar trend though with a narrower range. Clearly, the elephant's preference for grass leaves during the first wet season supplied it with over 8% protein. But they avoid the leaves with high protein (7–8%) and switch over to the bases (2–4%) during the second wet season. This could be for two reasons: the unpalatable nature of the siliceous leaves, and the presence of soluble carbohydrates in the more succulent basal portion (Field 1971, 1976).

During the dry season the protein level in the grass bases falls below 2.5%, which is insufficient for maintenance. Browse plants have high crude protein levels even in the dry period. The range of values is 8–10% in the Malvales and 10–20% in the Leguminosae such as *Acacia* and *Albizia*. It is not known how much browse protein is unavailable owing to interference by secondary compounds. Elephants showed a clear preference for browse in the dry season.

Elephant feeding on bark is not yet fully understood. Laws *et al.* (1975) suggested that, during the wet season, supplementing a diet of low-fibre grass with bark helps to maintain an optimum fibre:protein ratio for a correct throughput rate to ensure proper digestion of protein. The fibre in bark may help elephants avoid colic, to which they are prone (Eltringham 1982; also post mortem observations in elephants in southern India). Elephants in the study area, however, did not debark trees to a greater extent during the wet season, as was observed by Laws *et al.* (1975). Considerable amounts of bark were consumed during the dry season. McCullagh (1973) made an interesting observation that feeding on bark is a response by African elephants to a deficiency in essential fatty acids, such as linoleic acid, found in higher amounts in bark. The high content of minerals such as manganese, iron, boron, copper (Dougall, Drysdale & Glover 1964) and calcium (Bax & Sheldrick 1963; Laws *et al.* 1975) has also been suggested as the reason for feeding on bark, although one study found no relationship between degree of debarking and mineral content of plants (Anderson & Walker 1974).

As could be expected, the calcium content of dicotyledons' bark (range

7 mg/g) was much higher than that of grasses (1–5 mg/g). Although a diet
 ly grass could provide a sufficient intake of calcium for elephants, it is
not known how much of the calcium is physiologically unavailable. It is also
not necessary for an animal to be deficient in a particular nutrient for it to
consume foods rich in that nutrient. Supplementing the diet with bark could
certainly increase its calcium intake to a safe level. Bark may serve more than
one purpose in elephant nutrition.

Elephants are known to prefer water and soils rich in sodium (Weir 1973). All the wild plants analysed had a relatively low sodium content, although (as will be discussed later) certain cultivated crops had significantly higher amounts. Intestinal microbes can synthesize Vitamin B_{12} if cobalt is present in the diet. Plants belonging to the Palmae are good sources of cobalt (Ananthasubramaniam 1980). The role of other minerals in diet selection is not known.

5.6.3 Plant secondary compounds

Herbivores must not only obtain necessary nutrients from plants but also avoid the negative effects of plant defences, either physical or chemical. Physical defences such as thorns or spines in *Acacia* do not deter elephants from feeding. Chemical defences in the form of 'secondary plant compounds' are of great significance in herbivory (Freeland & Janzen 1974; Rosenthal & Janzen 1979). These have been broadly classified under two categories: quantitative defences or digestibility-reducing compounds, chiefly tannins, and qualitative defences or toxins such as alkaloids and cyanogenic glycosides. Herbivores can detoxify secondary compounds either by microbial enzymes in the gut or by microsomal enzymes in the liver.

Among the food plants of the elephant, the various species of *Acacia* have a high tannin content. The bark of *A. pennata* contains 9% tannin, and that of *A. leucophloea* may be up to 21% tannin (Wealth of India 1948); both are commonly consumed by elephants. Further studies are needed to see whether the degree of debarking and the tannin content are seasonally related. Presence of latex in plants of Moraceae and Anacardiaceae does not prevent elephants from feeding on them (also see Olivier 1978*b*). It is well known that hydrogen cyanide (HCN) concentrations are highest in immature tissues of plants, including the families Gramineae and Leguminosae (Conn 1979). Elephants perhaps avoid toxicity by feeding on these plants only after they have grown sufficiently for HCN levels to be negligible.

Elephants may also have to increase their dietary diversity in order to avoid excessive intake of any particular toxin in a plant, even though this would expose them to a variety of secondary compounds (Freeland & Janzen 1974).

5.6.4 *Browse and grass in the optimal diet*

There has been considerable difference of opinion as to whether elephants are primarily browsers or grazers. Based on the high proportion of grass in the diet (Buss 1961; McKay 1973) or the high preference for grass (Olivier 1978b) it has been suggested that elephants are mainly grazers. Others have stressed the need for consuming a higher proportion of browse plants (see, for example, Sikes 1971; Laws *et al.* 1975).

Grass is available in abundance in many regions, needs little time for 'preparation' before consumption and has low levels of chemical defences. It would be certainly advantageous to consume grass in bulk quantities during the wet season when it has sufficient protein for maintenance. But tall grasses such as *Imperata cylindrica* and species of *Themeda*, which are widespread in elephant habitat, are also very low in protein during the dry season. Elephant populations forced to live on a predominantly grass diet throughout the year suffer nutritionally to the extent that their fertility may decrease (Laws *et al.* 1975). An exception may be grasses in swampy areas such as the *villus* in Sri Lanka which may maintain a high nutritive value even during the dry season.

On the other hand, numerous studies have shown that browse is important during the dry season (see, for example, Field 1971; Field & Ross 1976; Guy 1976; this study). The African forest elephant is also predominantly a browser. Evidence from carbon isotope ratios in bone collagen from elephants in the study area indicates that 55–83% of the organic carbon in adults comes from C_3 plants (Sukumar *et al.* 1987, and unpublished results). The C_3 plants, including most dicotyledonous browse and bamboos, are not consumed to such an extent quantitatively. This means that elephants must be deriving a higher amount of organic nutrients, chiefly protein, from C_3 plants than from the C_4 plants (most tall grasses except bamboos) per unit quantity consumed.

Browse plants and their parts would, of course, vary considerably in nutritive value. The foliage of most leguminous plants has a high protein value. Bark is low in protein. Rain forest plants also have high levels of secondary compounds which would make them unpalatable to herbivores. This would explain the preference for grass in Malaysian forests. Ultimately, the optimal diet of elephants seems to be a 'proper' mix of browse and grass, depending on the season and the vegetation type.

6

Impact on the vegetation and carrying capacity

A central concept in population ecology is that of the 'carrying capacity', defined as the biomass or density of one or more species that a given area can support with its resources of space, food, etc. Much of the early theoretical framework, based on the logistic equation, was developed and tested using organisms growing under laboratory conditions. In the experimental situation, the carrying capacity is considered to be the density of a species when the population growth rate is zero. In nature the situation is much more complicated.

Our understanding of plant–herbivore systems lags far behind that of predator–prey or host–parasite systems (Caughley & Lawton 1981). Superficially it may appear that there is an abundance of plant resources as food for herbivores. In most terrestrial ecosystems the herbivores consume only 2–20% of the net primary production. This led to the postulation that the herbivore trophic level is usually not limited by food supply (Hairston, Smith & Slobodkin 1960; Slobodkin, Smith & Hairston 1967), a reasoning which has been questioned (Murdoch 1966; Ehrlich & Birch 1967). Studies on the Serengeti ecosystem have shown that 30–60% of the primary production is consumed by herbivores and nutritious food is indeed a limiting factor (Sinclair 1975).

For a large herbivore, in which population regulation through natural predation can be ruled out, it is all the more likely that food is a limiting resource. The regulation of elephant populations is of particular concern because of their potential impact on the vegetation. African elephants push over trees on a large scale and convert woodlands into grasslands in certain regions (Buechner & Dawkins 1961; Lamprey *et al.* 1967; Laws 1970; Wing & Buss 1970; Douglas-Hamilton 1972; Laws *et al.* 1975; Leuthold 1977; Barnes 1980). The 'elephant problem' has been a subject of intense and often emotional debate. One view is that destruction of trees on such a scale is unacceptable and culling is necessary to maintain a healthy elephant population and habitat status. The other view favours a *laissez-faire* policy of letting a natural ecosystem regulate itself with no human interference.

It is in this context that food resource production and consumption, impact of the elephant on the habitat and 'carrying capacity' have to be examined for the Asian elephant. The investigations I made on these aspects are preliminary. Long-term studies are needed to provide precise answers.

6.1 Primary production of grasses

In order to estimate the herbivore biomass that could be sustained during adverse years, it is prudent to measure primary production during such periods. Without such intention the estimates of grass production in the study area were made during 1982, which happened to be a period of drought.

The standing biomass of grass at two-month intervals at two study sites is shown in Fig. 6.1. There was no grazing by cattle or by elephant in Site 1 during the study period. In Site 2 the estimation was more complicated owing to erratic rainfall and some grazing by cattle. The relation between grass production and rainfall is given in Table 6.1. From these figures the net primary production of the grass-dominated herb layer in a deciduous forest can be taken as 2.5 g/m^2 for every 1 cm rainfall. Net primary production in semi-arid climates shows a nearly linear relation with precipitation (Whittaker 1970; Phillipson 1975; Sinclair 1975).

Table 6.1. *Relationship between grass production and rainfall*

Locality	Grass production (g/m^2)	Rainfall (cm)	Grass production in relation to rainfall (g/m^2 per cm rain)
Site 1 (Zone 3)			
July–August 1982	62	23	2.7
September–October 1982	63	25	2.5
Site 2 (Zone 16)			
July–August 1982	0	1	—
September–October 1982	36	15	2.4
Bandipur National Park			
April–November 1975	500[a]	197	2.5

[a] The grass production in Bandipur is based on Prasad & Sharatchandra (1984).

88   *Impact on vegetation and carrying capacity*

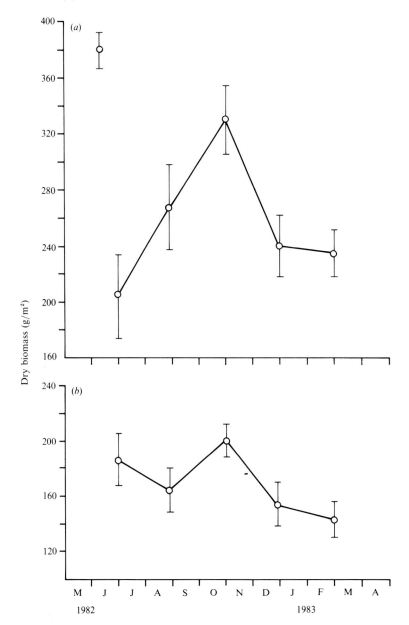

Fig. 6.1. Mean biomass of grasses at two study sites: (a) site 1, zone 3; (b) site 2, zone 16. Bars indicate ± one standard error.

Based on the above relation, the net primary production of grasses was estimated to be 195 g/m² in Site 1 (Zone 3) and 87 g/m² in Site 2 (Zone 16) during 1982. By making a correction for the area within a zone under actual grass cover, the average production was estimated to be about 175 g/m² in Zone 3 and 70 g/m² in Zone 16. Similar estimates based on the rainfall–production relation gave approximate figures of 100 g/m² for Zone 6 and 50 g/m² for Zone 11B during 1982.

Apart from rainfall, the productivity of grasses is also influenced by soil type, light, grazing and fire. Moderate grazing pressure may actually increase productivity, although it would reduce the standing biomass of grass at any given time (McNaughton 1976, 1979). Fire also stimulates grass productivity (e.g. Edroma 1984). In both the study sites there was no fire during the dry season of 1982 and thus considerable biomass was carried over from the previous growth season. The estimated production in these sites should be taken as the minimum levels for unburnt areas during a year of low rainfall. During years of normal or excess rainfall and widespread fire, the net primary production may be up to four times higher.

6.2 Biomass of mammalian herbivores and grass consumption

The biomass of large mammalian herbivores, both wild and domestic, is given in Table 6.2. A distinction must be made between 'crude biomass' and 'ecological biomass' (Eisenberg & Seidensticker 1976). Ecological

Table 6.2. *Biomass of large mammalian herbivores in the study area*

Species	Population range (within 928 km²)	Mean body weight (kg)	Crude biomass (kg/km²)	Food requirement (kg/individual/day)
Domestic				
Cattle	18 000–25 000	175	3394–4714	4.38
Buffalo	1700–2000	275	504–593	6.88
Sheep	6000–7000	40	259–302	1.00
Wild				
Gaur	100–200	600	65–130	15.00
Elephant	365–650	1800	708–1260	27.00
Spotted deer	600–900	45	29–44	1.35
Sambar	500–1000	145	78–156	4.35
Muntjac	300–600	20	6–13	0.60
Blackbuck[a]	400–600	25	11–16	0.75
			5054–7228	

[a] Found only in the Moyar valley at an ecological density of 6.5–10/km².

biomass within specific habitats may be several times higher than the crude biomass averaged over a larger area with different habitat types. The range of crude biomass of wild herbivores was 894–1619 kg/km^2 and that of domestic livestock was 4157–5609 kg/km^2 during 1982.

A comparison of crude herbivore biomass in Asian habitats is provided in Table 6.3. Other deciduous forest habitats such as Gir and Kanha, which do not have elephants, maintain a much lower wild herbivore biomass than the study area, although the combined wild and domestic herbivore biomass levels are comparable. The estimate of 8000 kg/km^2 for Bandipur is the ecological biomass within a restricted area. The wild herbivore biomass over the entire Bandipur National Park (area 874 km^2) may average only half this figure. Moister localities such as Chitawan and Kaziranga are capable of sustaining a much higher total biomass level.

The production and consumption of grass are given in Table 6.4 for four zones in the study area as examples of the tremendous variation in herbivore impact even between adjacent habitats in a region. In Zone 3, where grass production was high and livestock grazing negligible, only 6–7% of the production was consumed. By contrast, in Zone 11B over 50% of the production was utilized, primarily by domestic livestock. During a year of normal rainfall and about twice the primary production of 1982, the proportion consumed would be correspondingly lower (3–25%). The

Table 6.3. *Comparison of crude biomass of mammalian herbivores in Asian habitats*

Region	Area (km^2)	Crude biomass of mammals	
		Wild (kg/km^2)	Domestic (kg/km^2)
1. Satyamangalam–Chamarajanagar, India	928	1257	4883
2. Bandipur, India	20	8063	—
3. Kanha, India	319	667	4678
4. Gir, India	1166	383	6171
5. Kaziranga, India	430	2858	?
6. Chitawan, Nepal	545	1790	28 076
7. Wilpattu, Sri Lanka	580	766	?
8. Gal Oya, Sri Lanka	250	847	?

Sources of information: 1. This study (Table 8.1). 2. Modified from Johnsingh (1980). 3. Schaller (1967). 4–8. Eisenberg & Seidensticker (1976), based on other sources.

6.3 Impact on woody vegetation

The elephant's impact on *Kydia calycina, Grewia tiliaefolia, Acacia leucophloea* and *Acacia suma* was assessed by cross-sectional sampling during 1982. During subsequent years the utilization patterns were only qualitatively observed; a second sample for *A. suma* was taken in 1987. Demographic trends can be understood only by long-term monitoring of marked trees. This was not possible during the study, but the cross-sectional data provided some useful information.

(*a*) Kydia calycina *and* Grewia tiliaefolia

Figure 6.2 summarizes the intensity of utilization of *Kydia calycina* in Zones 3 and 15. The data for *Grewia tiliaefolia* are not presented but the pattern is similar to that for *Kydia calycina*, except that the density of the former is roughly one-third that of the latter.

Signs of feeding were seen in size classes up to 100 cm girth, although saplings below 10 cm were usually avoided. Damage to the 10–20 cm class represents the combined impact of elephants and gaur; above this size the utilization was almost exclusively by elephants. Feeding on the 10–40 cm girth class was both on branches and bark; damage to larger trees was restricted to stripping of bark.

Dead plants constituted less than 15% of plants of any size class. These were not always due to damage by elephants. Both *Kydia* and *Grewia* had a

Table 6.4. *Production and consumption of grass by herbivores*

All production and consumption values in tonnes/km^2.

Habitat	Net primary production	Consumption by mammalian herbivores			Percentage production consumed
		Elephants	Livestock	Others	
Zone 3	175	5.0	2.1	4.1	6.4
Zone 6	100	7.5	5.3	3.0	15.8
Zone 11B	50	2.2	21.1	0.5	47.6
Zone 16	70	13.2	21.1	2.3	52.3

Impact on vegetation and carrying capacity

good capacity for regenerating through stem coppices if the main stem was broken by elephants, or through root coppices if the stem was killed by fire. Their size class structure showed a healthy representation of seedlings and saplings below 10 cm girth. This indicates that their regeneration potential is high.

During the dry season of 1987, a very high concentration of elephants was

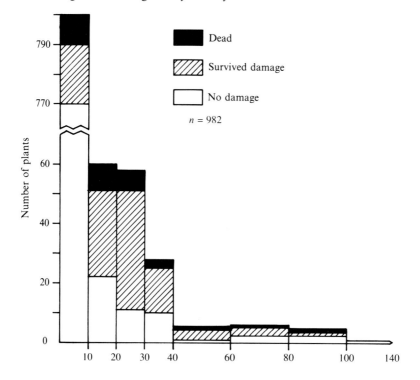

Fig. 6.2. Damage to *Kydia calycina*.

seen in the northern part of Zone 3. This was similar to the situation in 1983 (Chapter 4). Practically all ponds in the forests were dry in April 1987. Elephants stripped bark, largely from woody Malvales, on a scale never seen before in any region. In spite of the intensive utilization, the broken stems regenerated well after the rains.

(b) Acacia leucophloea

The proportion of *A. leucophloea* plants utilized in the 20–60 cm girth class was very high (Fig. 6.3). Over 90% of the trees in this class showed signs of feeding, mainly on the branches. Trees above 100 cm girth were immune to branch feeding and pushing over but some bark stripping was noticed. On the whole the mortality rate in *A. leucophloea* was negligible. Mortality in the mature tree class (>100 cm girth) was never due to elephant.

The size class structure shows a bimodal distribution with a near absence of trees in the 60–100 cm girth class. A similar bimodal size distribution has been described for *A. tortilis* in Lake Manyara National Park (Douglas-Hamilton 1972). The reasons for this are not clear but some speculations can be made. Owing to heavy browsing pressure, the younger trees may pass through a vulnerable stage from which only a few are recruited into the mature size class (Douglas-Hamilton 1972). There was no evidence for this in *A. leucophloea*, as no dead trees in the 60–100 cm class were found. During a past period corresponding to the age class of 60–100 cm girth, certain factors could have prevented seed germination or seedling establishment. It is possible that habitat manipulation through clearing of the undergrowth and setting fire occurred in the past. It is also likely that the *A. leucophloea* population is locked into a cyclic pattern of recruitment. Only trees above 100 cm girth were observed producing seed. Once this mature class dies out there could be a gap in seed production and regeneration until the younger trees grow into the mature class. Such a cycle could have been initiated through human activity, or it could be a response to long-term changes in environmental conditions as described for *A. xanthophloea* in the Amboseli basin (Western & van Praet 1973). The current regeneration was healthy.

(c) Acacia suma

A. suma shows a highly clumped distribution in Zones 7, 10 and 12 (Fig. 6.4). Damage was studied in one stand in Zone 12. Even though elephants utilized *A. suma* only during January–February, they made a considerable impact in 1981 and 1982.

All sizes above 10 cm were utilized intensively by elephants (Fig. 6.5). Breaking of the main stem was common. The most severe impact was on the

94 *Impact on vegetation and carrying capacity*

Fig. 6.3. Damage to *Acacia leucophloea*. Shading as in Fig. 6.2.; $n = 267$

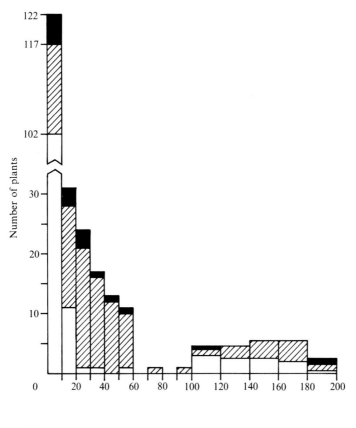

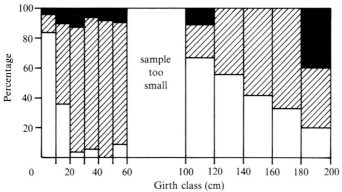

Impact on woody vegetation

20–40 cm girth class, in which about 80% of the trees were utilized and 50% were dead. The time period over which most of this utilization and mortality occurred was 1980–82.

Apart from the impact of elephants, the poor regeneration potential was of greater significance. The mortality in the girth class below 10 cm was largely not due to elephants. Possible reasons for the low regeneration could be failure of seed germination or edaphic conditions. Seed production by the mature trees was adequate.

In 1982 there clearly seemed to be a decline in *A. suma* woodland in Zone 12 from the short-term data. During subsequent years there was a sharp decline in utilization of *A. suma*. In 1983 there was very little damage since elephants occupied Zone 12 at a much lower density during the dry season than in previous years (Chapter 4). A second sampling of the stand in 1987 showed some interesting features (Fig. 6.5). The proportion of stems that had been utilized or were dead decreased noticeably in the above 30 cm girth class. In all size classes there was no sign of recent damage. Most of the damage or mortality seemed to have occurred prior to 1982. The size

Fig. 6.4. *Acacia suma* woodland (Zone 12) with trees broken by elephants.

structure of living trees, both utilized and not utilized, showed a pronounced shift towards the higher size classes due to growth of younger stems that had survived prior damage. There was a total absence of stems below 10 cm girth, indicating a high mortality of saplings or failure of seed germination. The density of *A. suma* stems was reduced but the population as a whole survived. Only a continued monitoring of the *A. suma* stand would reveal the extent to which elephant damage and edaphic or physiological factors govern its dynamics.

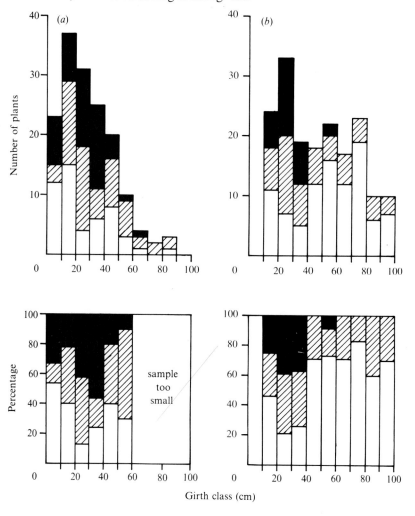

Fig. 6.5. Damage to *Acacia suma* assessed in (a) 1982, $n = 154$, and (b) 1987, $n = 175$. Shading as in Fig. 6.2.

(d) Plantation trees

Damage to *Eucalyptus* by elephants has been noticeable in recent times in many regions. Out of 602 trees marked in two 0.25 ha plots in Zone 16 during April 1982, 25 (4.2%) were broken by April 1983. Eucalyptus trees are harvested about 8 years after planting. Assuming no damage to young trees (below 2 years), about 22% of the planted trees are pushed over before harvest. The girth-class frequency distribution of a sample of 197 broken trees showed that 62% of these were between 15 and 30 cm.

Elephant damage to teak (*Tectona grandis*) and rubber (*Hevea brasiliensis*) plantations has also been reported from other regions of southern India.

6.4 Inter-specific competition, resource limitation and carrying capacity

To what extent is the elephant population limited by food supply and inter-specific competition? The availability of grass can be assessed from a balance sheet of primary production versus consumption by herbivores. This approach has been taken in 'grazing ecosystems' such as the Serengeti (see, for example, Sinclair 1975; Phillipson 1975). Estimating the productivity of shrubs and trees is a complex task. In such cases an assessment of whether the woody vegetation is regenerating or declining is the only approach possible (see, for example, Vesey-Fitzgerald 1973).

The apparent abundance of plant biomass can be misleading. The production taken as a mean value over the year is not always available as food for various reasons. Many plants are nutritious only during certain seasons. Grasses have a high protein content (8–10%) only during the wet season. During the dry season this value drops below 4%, which is insufficient for maintenance. The elephant is less affected by low-protein food than smaller herbivores are (Chapter 5). Plants or plant parts are high in secondary compounds during certain periods. Dry season fires which wipe out the coarse tall grasses would also destroy the more tender short grasses in localized areas. Such seasonal limitations in resource availability are important considerations which make annual figures of production irrelevant (Sinclair 1975).

The entire plant is not necessarily food. Elephants selectively consume only the top portion of grasses during the early wet season and later switch over to the basal portion. This feeding behaviour alone restricts availability to about 50% of the grass biomass at any given time. In woody plants an even smaller proportion is available for consumption. Only leaves and bark within reach of the elephant (below 5 m height) are available. Plants growing on steep, inaccessible slopes are largely out of reach for elephants.

Inter-specific competition may also be important in regulating animal numbers. An experiment conducted in the Mweya peninsula of Uganda illustrates inter-specific competition among herbivores (Eltringham 1974). Between 1957 and 1967 the number of hippopotamus was drastically reduced by culling. The most significant change during this period was a six-fold increase in the number of wild buffalo. This was attributed to an increase in the grass cover favouring the buffalo. Similar observations have been also made under natural conditions. In Serengeti the wildebeest and wild buffalo populations increased dramatically owing to an abundance of food induced by high rainfall during the dry season for a few years beginning in 1971. These two herbivores removed most of the extra food, preventing any significant increase in the populations of other grazers such as zebra, topi, kongoni and impala (Sinclair 1979).

Berwick (1976) found signs of severe overgrazing by domestic cattle and buffalo in the Gir Sanctuary, but concluded that this had no effect on the wild ungulates (spotted deer, sambar, four-horned antelope, chinkara, nilgai) which are mainly browsers in that region. In the study area the wild ungulates affected by livestock are probably the gaur in the tall grass habitats and the spotted deer in the short grass habitats. Cattle and buffalo could be expected to compete with gaur for high-quality tall grass, which is scarce during the dry season. The spotted deer spends between 68% (dry season) and 95% (wet season) of its feeding time in grazing at Bandipur (Prasad & Sharatchandra 1984).

Is there competition between livestock and elephants? From available data, it would seem that elephants, being relatively coarse feeders, are not seriously affected by the consumption of tall grass by livestock. A comparison between the production and consumption levels in the study area (Table 6.4) shows that, in Zone 3 and Zone 11B, elephants managed an offtake of only 3–4% of the primary grass production irrespective of the levels of consumption by livestock.

On the whole, it is difficult to imagine that grass is a limiting factor for elephants in the deciduous habitats. Because the elephant is a mixed feeder it is more likely to be affected by the availability of high-quality browse during the dry season. If grass production is in excess of available browse production, and if elephants have to consume browse in at least equal amount to grass in order to maintain condition (Laws *et al.* 1975; Chapter 5), the 'excess' grass production will go unutilized, assuming that the population is regulated by browse production. In deciduous habitats many shrubs and trees shed their leaves during the dry period. Elephants have to meet their protein requirement from green foliage and not merely from twigs or bark. Generally, some edible browse plants are in leaf at any given time, but this may not be true during an extended dry spell. For instance, in March–April 1983 the shrubs *Acacia pennata* and *A. torta*, which make an important contribution to the dry-season diet, were almost totally defoliate. A detailed study on the phenology of browse plants is necessary before their role in limiting food supply can be determined.

There was no evidence that elephant utilization of woody vegetation in the study area would result in a decline of tree populations. The woody Malvales coppiced well even if the stems were broken. Factors other than elephant damage also seemed important in the dynamics of *Acacia leucophloea* and *A. suma*. The shift in seasonal centres of high elephant density over a time period, due to fluctuations in annual rainfall, may allow tree populations to recover from localized intensive damage. Fire is also an important factor regulating the regeneration and survival of woody plants (Chapter 9).

Water may be an even more important limiting resource for elephants. Hanks (1979) observed that the stomach and intestinal contents were abnormally dry in elephants which died during a drought in 1970 in the Siloana plains of Zambia. In the study area, elephants depend to a large extent on stagnant water from small ponds or streams. The risk of dehydration or contracting disease from contaminated water may increase. There was, however, no evidence for an increased risk of natural mortality during the dry months compared with other times (Chapter 12).

6.5 Elephant–vegetation interaction and ecosystem dynamics

There has been intense debate regarding the impact of *Loxodonta* on the habitat. One school considers woodland destruction as artificial and unacceptable (see, for example, Laws 1970; Buss 1977) while the other relates habitat change to long-term natural ecosystem processes (see, for example, Caughley 1976; Norton-Griffiths 1979).

The 'compression hypothesis' has been stated in various forms by Buechner & Dawkins (1961), Lamprey *et al.* (1967), Laws (1970) and Buss (1977). This hypothesis links habitat change to the elephant's overabundance due to compression of its range by expansion of human settlements. Elephants have no scope for dispersal from their 'island' habitats. In protected areas a reduction or elimination of hunting sets the stage for a rapid growth of the elephant population beyond the levels of the carrying capacity of the habitat. The rate of destruction of trees is unacceptable to the overall objectives of resource management. In its extreme form, it has been speculated that the elephant may even have contributed to the creation of deserts in Africa (Laws 1970). Because the 'elephant problem' has been created by human activities, it is irrational to expect that natural regulatory mechanisms can check the elephant population levels. The obvious solution, according to the supporters of this hypothesis, is an organized, scientific programme of culling the elephant population to a level at which its impact on the habitat would not cause deterioration (Pienaar 1969; Laws 1970). Such action would help maintain a high level of animal productivity, utilize rather than waste the secondary production, assure an adequate carrying capacity for other mammals and maintain a high biological diversity (Buss 1977).

In support of this reasoning, the experiences in the Tsavo and the Kruger National Parks are usually mentioned. In Tsavo the administrators decided to let nature take its course; the result was a high mortality of elephants in the park during the drought of 1970–71 (Corfield 1973). On the other hand, a culling programme has been operating since 1969 at Kruger. Here, the elephant population has been stable even during periods of drought, other rare mammals in the park have been holding their ground and the mature trees have not declined (Buss 1977; Hall-Martin 1981).

Not all studies on tree damage have implicated the elephant as the main culprit. Western & van Praet (1973) found that a rising water table and associated salinity created a physiological drought which killed a large proportion of *Acacia xanthophloea* trees in the Amboseli basin of Kenya. They presented evidence of long-term climatic cycles which caused corresponding changes in the composition of the flora and fauna. Croze (1974) argued that the regeneration potential was adequate to replace loss of *Acacia* trees from elephant damage in the Seronera area of Serengeti. He recommended fire control to enable the young *A. tortilis* to establish, but saw no need for culling elephants.

Whereas the earlier hypothesis assumed that under natural conditions the elephant–woodland system would exist in a stable equilibrium, Caughley (1976) began from the opposite view, that there is no natural equilibrium between elephants and forests in parts of Africa. He visualized a cyclical relationship in which elephants increased and began to thin out the forest, and later declined when the trees became sparse, thus allowing the forest to regenerate. This constitutes a 'stable limit cycle' in which the trends in tree and elephant densities are similar to sine waves, with the peaks or troughs of tree density being about one-quarter of a wavelength behind that of elephants. Based on the age structure of the baobab (*Adansonia digitata*) tree, he proposed that the expected period of the cycle was around 200 years in the Luangwa valley of Zambia.

The effects of time lag enable a large herbivore, which maintains a high biomass level, to exceed the carrying capacity (K) before density-dependent factors cause a decline in the population. Elephants further complicate matters by their ability to switch over from browse to grass if the vegetation changes owing to woodland destruction. There is evidence that density-dependent relations in large mammals are not linear (Fowler 1981). Such populations are most productive when close to K (and not at $K/2$ as implied by the logistic equation). Density-dependent brakes begin to operate only at levels close to K, but these do not prevent the population from overshooting the K level.

The hypothesis of Caughley (1976) was questioned on two counts (see, for example, Hanks 1979). Firstly, there is no historical evidence for a low elephant density and a devastated forest in a state of recovery for Luangwa or any other part of Africa. Secondly, there is no evidence to support a direct relation between tree density and the fertility or mortality rate in elephants. Both these arguments are not necessarily true. The Tsavo National Park had a low elephant density during the late 19th century (Myers 1973). Some evidence for a decline in reproduction due to a predominance of nutritively poor grass in the diet, as a consequence of conversion of woodland into grassland, has been presented by Laws et al. (1975) for the elephants in the Murchison Falls Park of Uganda. Elephants in the Luangwa Valley have also shown a decline in fertility during 1980–81 (Lewis 1984), about a decade after a high fertility was recorded by Hanks (1972a). One of the suggestions given for this decline is a skewed sex ratio (1 adult male for 2.5 adult females). It seems unlikely that this alone could have caused the observed fertility decline (Chapter 12). Sikes (1968) further indicates that a loss of tree cover could increase elephant mortality from 'stress' diseases such as medial sclerosis and atheroma.

The massive elephant mortality in Tsavo (Corfield 1973), with its density-independent characteristics, also deviates in a sense from the concept of a stable limit cycle. Such smooth, symmetrical functions may never be encountered in nature. Caughley's (1976) model provides a conceptual framework in which to consider the oscillations in an ecosystem. Norton-Griffiths (1979) has suggested that the Serengeti ecosystem could oscillate between more-woodland and more-grassland phases. The possible scenario is as follows. In the primarily grassland phase, the high standing crop of grass results in intense fires which inhibit the establishment of woody vegetation, while promoting the growth of the grazing ungulates. The expanding grazer populations reduce the grass cover to levels where fires decrease in intensity owing to insufficient fuel. This, in turn, enables the woody plants to establish, decreases the extent of grassland and causes a decline in the grazer populations. Released from grazing pressure, the grass biomass once again creates the conditions for intense fires and a repeat of the cycle. During the past century, an exotic factor interacting with this system is the rinderpest virus, which has periodically decimated the populations of buffalo, wildebeest and other ungulates. In this model the browsers have not been considered. They would contribute to a more complex scenario.

Plant–herbivore interactions are compounded by genetic, climatic and spatial variability (Crawley 1983). Ecological thought has long been dominated by an improper understanding of 'stability'. Holling (1973) made out a difference between stability and resilience in an ecosystem. He defined stability as 'the ability of a system to return to an equilibrium state after a temporary disturbance; the more rapidly it returns and the less it fluctuates, the more stable it would be'. Resilience was defined as 'a measure of the persistence of systems and of their ability to absorb change and disturbance and still maintain the same relationships between populations or stable variables'. The implications of these definitions are clear. A system can have a low stability but a high resilience and *vice versa*. An example of a 'high stability – low resilience' system would be a climax rain forest, and one of a 'low stability – high resilience' system a desert or semi-arid ecosystem such as the East African savannas (Norton-Griffiths 1979).

Although numerous abiotic factors have to be considered, one obvious factor that influences the functioning of an ecosystem is rainfall. The higher the rainfall, the lower its annual variation. Evergreen forest is a climatic climax in regions of high rainfall. The coefficient of variation in rainfall is only 10–20% in regions bearing tropical evergreen rain forest in India. It is over 50% in the arid, xerophytic zones of the Thar desert.

With this appreciation of ecosystem stability–resilience, the characteristics of such systems can be considered with reference to elephant–vegetation interaction. Caughley (1976) suggested that in a stable limit cycle the period and amplitude should change along a climatic gradient, but he did not elaborate further. Based on the relation between rainfall and ecosystem stability–resilience, I suggest the following model for pristine habitats (Table 6.5, Fig. 6.6.).

(a) In a highly fluctuating semi-arid savanna woodland environment, the stable limit cycle of the elephant–vegetation system would possess a relatively short period and high amplitude. The range of elephant density would also be high. Reported values for East Africa are generally between 1 and 5 elephants/km^2 (Croze,

Hillman & Lang 1981). This ecosystem would witness a noticeable emergence and decline of elephants and trees.

(b) In a highly stable equatorial climax rain forest, the stable limit cycle would contract to a stable equilibrium. Elephants would exist at a low density. For the Malaysian rain forests, Olivier (1978b) estimated a density of about 0.1 elephant/km^2. The problem of elephant 'damage' to the woody vegetation would not arise in this habitat.

The factors contributing to this scenario could be as follows. It is clear that selection pressures would have varied across this continuum of stable to highly fluctuating environments, resulting in biotic components, both plants and animals, whose genomes are adapted to their specific environments.

Table 6.5. *Parameters of elephant–vegetation dynamics in different habitats*

	Semi-arid savanna woodland	Equatorial rain forest
Rainfall		
Total quantity/year	low (<70 cm)	high (>200 cm)
Coefficient of variation	high (>50%)	low (<20%)
Vegetation		
Density of large trees	low	high
Total biomass of vegetation	low	high
Proportional availability of woody plants as food	high	low
Proportion of total trees that can be pushed over by elephants	high	low
Quantity of edible woody plants	high	low
Carrying capacity, K		
Mean K (elephant/km^2)	high (>1)	low (<0.1)
Variation in K	high	low
Demography		
Age at first calving in female elephants		
Mean age in years	low (10–14)	high (>20)
Variance	high	low
Mortality rates	high range	low range
Maximum population growth rates per year	up to 4%	<1%
Elephant–woodland dynamics	highly fluctuating	stable equilibrium

Impact on vegetation and carrying capacity

Selection pressures could be expected to have moulded the genotype of *Elephas* in the dry regions of the Indian sub-continent somewhat differently from those in the rain forests of Malaysia. This difference may be subtle and seemingly non-existent (after all, an elephant is an elephant). Variations in social structure of a species from one habitat type to another are well known. Differences in selection pressures may also be reflected in the reproductive characteristics of a species.

In a highly fluctuating environment (semi-arid region) it would be adaptive for the elephant to have the capacity for a higher intrinsic rate of reproduction. It should be able to build up its numbers fast enough after a decline, either gradual or drastic, to avoid the possibility of extinction due to another perturbation. This would not be true for elephants in a stable environment (rain forest) where perturbations from the normal state are rare.

One possible reproductive parameter which could be moulded by natural selection is the age of sexual maturity (in this case, especially in females). This character could be a genetic trait subject to environmental modification (through temperature, nutrition, etc.). If this hypothesis were true, the age of sexual maturity should be lower in female elephants in an unstable environment than in those in a stable environment.

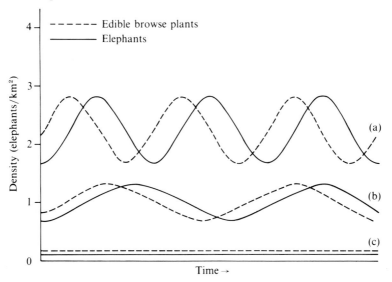

Fig. 6.6. Stable cycle model of elephant–vegetation dynamics in different environments: (a) savanna woodland, (b) deciduous forest and (c) equatorial rain forest.

The savanna woodland has a much higher carrying capacity for elephants than a rain forest. The proportion of total woody plant species available as food for elephants is high. The short-statured vegetation also ensures that a high proportion of the standing biomass can be consumed. The abundance of grass further contributes to a high carrying capacity. Crude densities reported for East Africa are between 1 and 5 elephants/km^2. In a climax rain forest the proportion of total woody species available as food is low, most of the biomass is trapped in large trees and unavailable for elephants, and grass is scarce (although the monocotyledonous palms substitute for grass to a limited extent). Elephant density is usually below 0.1/km^2, as in Malaysia.

The high elephant density and its impact on a large proportion of woody plants in a savanna woodland would result in a noticeable decline in trees, followed later by a decline in elephants. The decline in elephants could result from an increased calving interval owing to poor nutrition. The relatively high density of elephants and their ability to subsist on grass would ensure that the population does not cross the extinction threshold. The low age of sexual maturity and a progressively shortening calving interval owing to improved nutrition (from browse regeneration) would increase elephant numbers at a rate of up to 4% per year during later stages, resulting in damage to trees and repeat of the cycle. Elephant and tree densities would fluctuate with a high amplitude and a short period. In rain forests the elephants' impact on trees in the undergrowth would be hardly noticeable and a high proportion of large trees would anyhow be immune to being pushed over. Elephants and trees would be in stable equilibrium. On the scale of *r–K* selection, the elephant in the stable environment would be *K*-selected to a slightly higher degree than its counterpart in the fluctuating environment.

Deciduous forest, such as those in the study area, with 'medium' rainfall, carrying capacity and elephant density would be intermediate between the stable and fluctuating environments.

Evidence for or against the model can be obtained by examining elephant populations in a gradation of natural ecosystems. At present the necessary breadth of habitat types is not available for the Asian elephant. Manipulative research is also not possible. The African elephant encompasses the necessary range of climatic and vegetation types from equatorial rain forest to semi-arid savannas. Studies on the forest elephant (*Loxodonta africana cyclotis*) are especially important, as very little is known about its population biology.

The late age of sexual maturity (over 22 years) in female bush elephants in Budongo, Uganda (Laws *et al.* 1975) may be evidence for this model. The

Budongo forest is a moist semi-deciduous climax with an annual rainfall of 155–180 cm. On the other hand, an early age of maturity (12 years) has been recorded at Mkomazi and Tsavo, both of which are regions of low rainfall (38–58 cm in Mkomazi (Field 1971); below 50 cm in Tsavo (Myers 1973)).

Other models of elephant–woodland dynamics have also been proposed. Soil nutrient levels and water infiltration rates influence plant productivity and hence elephant densities and woodland destruction (Bell 1981; Botkin, Mellilo & Wu 1981). Thus even within semi-arid vegetation types the extent of the elephant problem would vary enormously. Jachmann & Bell (1984) have also suggested that the East African elephants causing destruction of woodland are descendants from elephants in the moist forests of Central Africa. After over-exploitation for ivory had greatly reduced elephant numbers in East Africa, there may have been migration from Central Africa. The feeding strategy of elephants which evolved in moist forests is maladapted for the savannas.

Such models may apply to large natural ecosystems, but human interference could substantially change the picture. Compression of elephants to abnormally high densities could upset natural regulatory mechanisms, resulting in a crash in numbers. Human-induced fires may suppress regeneration of woody plants and keep the system permanently in a grassland stage. Culling of elephants would reduce pressure on trees and impose an artificial equilibrium on the system.

What do the theoretical models and observed facts imply for management of elephant and other large mammal populations? The predictions of irreversible population disaster and desertification associated with African elephants, or other mammals, have not been proved as yet. For making accurate predictions of overabundance we need to know the location of extinction boundaries with different parameters (Sinclair 1981). The culling of mammals in North American parks is, in Sinclair's words, based on 'the ridiculous conclusion that the only good herbivore population is one vanishingly small'. With our current knowledge, or rather the lack of it, one approach may be to allow the perturbation experiments to continue and observe the behaviour of the system (Sinclair 1981). Decisions have also to be made on socio-economic compulsions in a region. In Kenya, unlike in north America, a park should 'sustain not only the spirit but also the stomach' (Myers 1973). The thousands of elephant carcasses in Tsavo, left to the vultures, could have been a protein source for the 180 000 people living on famine relief if a properly organized programme of culling had been in operation prior to the population crash.

The problem of destruction of forest would not arise for a large area of the Asian elephant's range, covering the moist forests. In the drier regions this problem is not anywhere as serious as in Africa. There may be a behavioural difference between *Elephas* and *Loxodonta* with respect to tree destruction. Because female *Elephas* do not possess tusks they may cause less damage to trees. Also *Elephas* do not seem to push down trees with the same propensity as *Loxodonta*. The cycle of elephant–tree interaction could have a lower amplitude and longer period in the Asian habitat than in a similar African habitat.

There is no justification at present for culling Asian elephants on the grounds of habitat destruction. Even if the problem arises in regions of high elephant density such as southern India or Sri Lanka, the cultural traditions here would rule out the killing of elephants. The alternative to a *laissez-faire* policy would be capture of elephants for domestication.

7

Crop raiding by elephants

Depredation of crops by elephants occurs to varying extents throughout their present range in Africa and Asia, wherever cultivation abuts elephant habitat. In spite of the extensive research on the African elephant (*Loxodonta africana*), the interaction between elephants and agriculture has, surprisingly, received little attention from an ecological perspective. The work of Allaway (1979) is in his own words 'more descriptive than analytical' and contains no quantitative data on crop raiding. Studies on the Asian elephant (*Elephas maximus*) only state the problem briefly (McKay 1973; Olivier 1978b) or give the economic implications of crop damage (Mishra 1971; Blair *et al.* 1979; Blair 1980). The present study considers crop raiding behaviour in relation to the elephant's life-history strategy (Sukumar 1989a,b).

7.1 The crops cultivated

The locations of the study villages are shown in Fig. 7.1. Agriculture in the region is mainly traditional rain-fed cultivation (Fig. 7.2). The perennial or semi-perennial plantation crops include coconut (*Cocos nucifera*), banana (*Musa paradisiaca*) and sugar cane (*Saccharum officinarum*), but these are grown on a small scale. Jackfruit (*Artocarpus integrifolia*) and mango (*Mangifera indica*) trees are found in some gardens. The millets, cereals, pulses and oilseeds are cultivated in two fairly distinct crop seasons (Fig. 7.3).

(a) Minor crop season: The cultivation of sorghum (*Sorghum vulgare*) and maize (*Zea mays*) begins after pre-monsoon showers during April–May. These rely mainly on the south-west monsoon (first wet season) during June–July and the crop is ready for harvest by August. Only a small fraction (4%) of the cultivable land area was cultivated for these crops during this season in 1981. Gingelly (*Sesamum indicum*) is also grown during this period.

The crops cultivated

Fig. 7.1. Map showing location of the study villages. The 12 villages shown in capitals were regularly monitored for crop damage. (From Sukumar 1989b.)

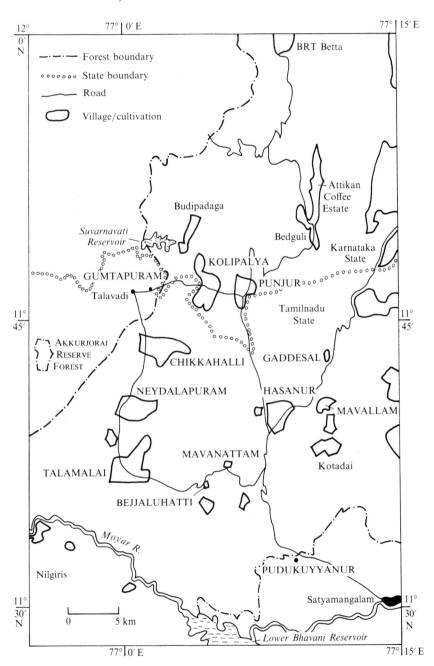

110   *Crop raiding by elephants*

Fig. 7.2. A panoramic view of cultivated fields in Hasanur, the focal study village, facing westwards.

Fig. 7.3. Proportion of land under cultivation of crops edible to elephants. (From Sukumar 1989b.)

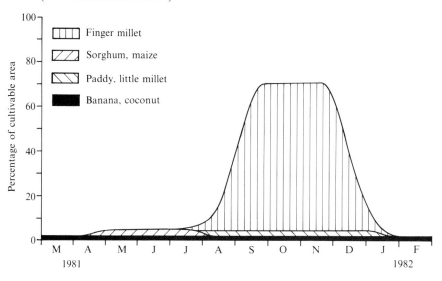

(b) Major crop season: The staple food crop of the people, *ragi* or finger millet (*Eleusine coracana*) is usually sown during August–September before the end of the southwest monsoon. It depends for its growth mainly on the northeast monsoon (second wet season) during October–November and is ready for harvest in December–January. Most farmers also plant rows of niger (*Guizotia abyssinica*), an oilseed crop, along with finger millet. In 1981 finger millet was cultivated over 65% of the land area. Paddy (*Oryza sativa*) and little millet (*Panicum miliare*) were grown to a far lesser extent, occupying 1.5% and 0.5% of the area respectively. Among the legumes, horse gram (*Dolichos biflorus*) is cultivated mixed with other crops.

7.2 The crops consumed and patterns of feeding

Almost invariably elephants entered settlements only after it was dark. During the crop season, adult male elephants would usually move close to the forest–village boundary in the evening and cross over into the fields after sunset. Bulls were often seen entering between 19.00 hr and 22.00 hr and leaving before sunrise. There were some exceptional cases of both bulls and family herds straying into cultivation during the day. In the case of herds it was because farmers had chased them away at night from their land bordering the forests and the elephants had run in confusion further into cultivation instead of going back into the forest.

Once elephants enter cultivation their next goal is to find a suitable crop field for foraging. Finger millet made the highest contribution among crops to the elephants' diet in the study area (Fig. 7.4). During this season paddy and little millet fields were also visited, especially when these were harvested after the finger millet crop. There was strong selection for fields with robust plant growth, high density of standing crop and plants in inflorescence or grain stage. Elephants rarely consumed finger millet plants in vegetative stage from fields they had to cross before reaching the target field. They also invariably consumed only the terminal portion of the plant bearing the inflorescence, while discarding the basal stem with the roots. On an average the terminal portion consumed formed 62% (s.d. $= 10.5, n = 31$ samples of 30 plants each) by weight of the finger millet plant. When feeding on sorghum and maize, elephants are again selective in plucking the stalk bearing the inflorescence. In the absence of spikes they may eat the terminal portion of the succulent stem and the leaves.

From young coconut trees, below 2 or 3 m in height, only the central rachis is pulled out and eaten (Fig. 7.5). Such small trees are highly vulnerable to damage (Table 7.1). Taller trees are uprooted, usually by adult bull elephants, before the rachis is plucked. Banana stems are split and the fibrous

112 *Crop raiding by elephants*

Fig. 7.4. Finger millet (*Eleusine coracana*) field with damage by elephants in the foreground.

The crops consumed and feeding patterns

Fig. 7.5. Coconut trees damaged by elephants. The central rachis has been pulled out from the trees.

Table 7.1. *Size class frequency of coconut trees damaged by elephants*

Height class of trees (metres)	Number of trees damaged	Percentage
Below 1 m	103	67.3
1–2	25	16.3
2–3	8	5.2
3–4	6	3.9
4–5	6	3.9
Above 5 m	5	3.3
Total	153	

pith consumed, and also occasionally the inflorescence spike or fruit bunch. Sugar canes are broken, chewed and consumed whole. A preference for mango trees was noticed from the time of flowering until fruiting, such shoots being selectively eaten. Jackfruits are plucked or shaken off the trees, crushed with the foot and consumed.

7.3 Frequency and seasonality of raiding

Once a particular field or cluster of fields was selected, an adult bull continued raiding it for a few consecutive nights before turning its attention to another location. Although fields bordering the forest usually bore the brunt of the damage, a more central location within an enclave was no immunity to attack by bulls. Within cultivated tracts the bulls moved freely, often walking a distance of up to 6 km across the tract. Elephant herds largely confined their forays to 1 km from the forest boundary, although there were a few instances of them traversing the large tract of Kolipalya village. Raiding of a particular field was less predictable in the case of herds.

An adult bull also raided crops far more frequently than did a member of a family herd (Fig. 7.6). Throughout the year, an average adult bull raided fields on 49 days, whereas an average member of a herd did so on only 8 days. One notorious bull (MA-6) entered cultivation at least 120 days in the year.

The frequency of crop raiding in the villages during different months (Fig. 7.7) was, as could be expected, proportional to the area of land under cultivation. During the dry period from February to April, when most of the land was in fallow, raiding was sporadic. Solitary bulls made occasional incursions to feed on plantation crops such as coconut, banana, mango, jackfruit and sugar cane. These raids were mainly in the villages of

Frequency and seasonality

Fig. 7.6. Frequency of crop raiding by elephants.

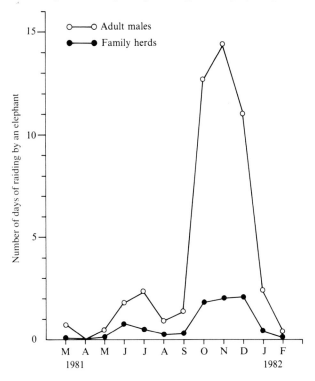

Fig. 7.7. Frequency of crop raiding by males (M) and family herds (F) in the study villages. (From Sukumar 1989b.)

Village	Mar M	Mar F	Apr M	Apr F	May M	May F	Jun M	Jun F	Jul M	Jul F	Aug M	Aug F	Sep M	Sep F	Oct M	Oct F	Nov M	Nov F	Dec M	Dec F	Jan M	Jan F	Feb M	Feb F
Hasanur	0	0	0	0	1–5	0	6–10	0	6–10	0	0	0	6–10	0	>26	0	>26	0	>26	6–10	11–15	1–5	0	0
Punjur	0	0	0	0	0	0	1–5	1–5	0	1–5	0	0	6–10	6–10	21–25	21–25	21–25	21–25	11–15	21–25	0	0	0	1–5
Kolipalya	0	0	0	0	0	0	1–5	1–5	0	0	0	1–5	0	1–5	>26	21–25	11–15	21–25	6–10	6–10	0	0	0	0
Gumtapuram	0	0	1–5	0	1–5	1–5	1–5	6–10																
Chikkahalli	1–5	1–5	0	0	0	0	1–5	1–5	0	0	0	0	0	1–5	>26	6–10	6–10	6–10	6–10	11–15	1–5	1–5	0	0
Neydalapuram	1–5	0	0	0	0	0	1–5	0	1–5	1–5	0	0	1–5	0	1–5	0	11–15	11–15	21–25	11–15	11–15	1–5	0	0
Talamalai	1–5	0	0	0	0	1–5	0	0	1–5	0	1–5	0	0	0	6–10	6–10	21–25	21–25	21–25	11–15	6–10	6–10	6–10	1–5
Bejjaluhatti	0	0	0	0	0	0	0	0	0	0	0	0	0	0	0	0	0	0	0	0	0	0	0	0
Mavanattam	0	0	0	0	0	0	0	0	0	0	0	0	0	0	0	0	0	0	1–5	1–5	11–15	1–5	0	0
Gaddesal	0	0	0	0	0	0	1–5	0	1–5	0	21–25	0	0	0	0	0	0	0	0	0	0	0	0	0
Mavallam	0	0	0	0	0	0	0	0	0	0	0	0	0	0	0	0	0	0	1–5	0	1–5	1–5	0	0
Pudukuyyanur	0	0	0	0	0	0	0	0	0	0	0	0	0	0	0	0	0	0	1–5	1–5	6–10	1–5	0	0

Number of days of raiding per month: ○ 0, ◐ 1–5, ◔ 6–10, ● >26, ◕ 11–15, ◑ 16–20, ◗ 21–25

Chikkahalli, Neydalapuram and Talamalai. From May to August, a few bulls and herds operated in the northern villages of Punjur, Kolipalya and Gumtapuram, where some farmers cultivated sorghum and maize. During this season, raiding was at a relatively low frequency of less than 10 days in a month. Solitary bulls, however, raided maize fields more often at Gaddesal, where a large proportion of this small settlement was cultivated.

The frequency of raiding reached a peak during the finger millet crop season when a large area was cultivated. Some villages were raided practically every night. Intensive raiding by both bulls and herds began during the last week of September when finger millet flowered in the fields of Punjur and Kolipalya, and continued until the harvest began in December. Hasanur was raided mostly by bulls from late September until mid-January with only a single herd raiding on consecutive nights in sub-groups during the first week of December before moving in a southwest direction.

On the western sector at Chikkahalli, raiding was at a peak during October by mainly bulls and, to a lesser extent, by herds. Raiding gathered momentum in November at Neydalapuram and Talamalai, continued at a high frequency during December and decreased in January. Sporadic raiding occurred at Talamalai in February, mostly by bulls and twice by herds for the harvested finger millet. The small settlement at Mavanattam was visited in late November and early December by some adult bulls and a couple of small herds. Nearby Bejjaluhatti did not experience any raiding.

Gaddesal did not cultivate finger millet in 1981. Mavallam was visited only a few times late in the season by adult bulls. In the plains to the south, raiding by bulls and herds began in the bordering villages including Pudukuyyanur in late November or early December, but was generally at a low frequency. There was no raiding between March and October in this region.

7.4   Group sizes of raiding elephants
7.4.1   *Mean monthly raiding sizes of bulls and herds in villages*

Tables 7.2 and 7.3 give the mean raiding sizes or number of adult bulls and elephants in family herds visiting a village in a day during each month. While raids by just one bull per day were the most common in all the villages, up to 6 bulls were recorded on certain days during October–November at Hasanur. In all the villages the raiding sizes of bulls were highest during the finger millet season. The largest family herd sizes recorded were in the village of Kolipalya during June–July and October. Here between 15 and 25 elephants raided the fields on many occasions. In general, sizes of raiding herds were between 3 and 10 elephants with a mean size of 7.9 for all villages.

Table 7.2. Mean raiding group size of adult bulls

Village	1981											1982	
	Mar.	Apr.	May	June	July	Aug.	Sept.	Oct.	Nov.	Dec.	Jan.	Feb.
Hasanur	0	0	1.0	1.0	1.4	0	2.2	3.4	4.5	3.1	1.0	0
Punjur	0	0	0	1.0*	0	0	1.0	2.4	1.5	1.9	0	0
Kolipalya	0	0	0	1.0*	0	0	1.0*	2.1	2.3	2.0	0	0
Gumtapuram	0	1.0	1.3	1.0	—	—	—	—	—	—	—	—
Chikkahalli	1.0	0	0	1.0*	0	0	1.0*	2.3	2.0	1.0*	1.0	0
Neydalapuram	1.0	0	1.0	1.0	1.0	0	1.0*	1.0*	1.5	2.0	2.3	0
Talamalai	1.0	0	1.0*	0	1.0*	0	0	1.5	1.3	1.3	1.0	1.0
Bejjaluhatti	0	0	0	0	0	0	0	0	0	0	0	0
Mavanattam	0	0	0	0	0	0	0	0	1.0*	1.0	0	0
Gaddesal	0	0	0	1.0	1.5	2.0	0	0	0	0	0	0
Mavallam	0	0	0	0	0	0	0	0	0	1.0*	2.0	0
Pudukuyyanur	0	0	0	0	0	0	0	0	0	1.0*	1.0	1.0*

The total sample size is 258 days of raiding.
* Based on only one instance of raiding during the month.
— Data not gathered.
From Sukumar (1989b).

Table 7.3. Mean raiding group size of family herds

Village	1981											1982	
	Mar.	Apr.	May	June	July	Aug.	Sept.	Oct.	Nov.	Dec.	Jan.	Feb.
Hasanur	0	0	0	0	0	0	0	0	0	9.5	4.0*	0
Punjur	0	0	0	5.0	7.0	0	8.5	7.0	6.8	6.8	0	8.0*
Kolipalya	0	0	0	15.7	15.0*	4.7	6.5	14.3	6.6	4.0	0	0
Gumtapuram	0	0	5.0	11.5	—	—	—	—	—	—	—	—
Chikkahalli	4.0*	0	0	0	0	0	0	4.0	7.5	7.5	12.5	0
Neydalapuram	0	0	0	0	7.0	0	0	0	8.8	8.4	3.0	0
Talamalai	0	0	0	0	0	10.0*	0	6.0	6.4	13.3	6.5	4.0*
Bejjaluhatti	0	0	0	0	0	0	0	0	0	0	0	0
Mavanattam	0	0	0	0	0	0	0	0	5.0	4.0*	0	0
Gaddesal	0	0	0	0	0	0	0	0	0	0	0	0
Mavallam	0	0	0	0	0	0	0	0	0	0	3.0*	0
Pudukuyyanur	0	0	0	0	0	0	0	0	0	4.3	6.2	0

The total sample size is 120 days of raiding.
* Based on only one instance of raiding during the month.
— Data not gathered.
From Sukumar (1989b).

7.4.2 Frequency of unit group sizes

In contrast to the total number of elephants involved in raiding villages, the frequency of group sizes or the number of elephants per unit bull or herd group showed some interesting patterns. Adult bulls showed a distinct tendency to associate in larger groups while raiding at night, compared with their normal association in the forest during the day (Table 7.4). Groups of up to 4 bulls coming together for raiding were recorded. The general pattern of male–male association during raids was that bulls raiding solitarily were usually over 25 years old (in 22 out of 24 cases where age was determined from footprint circumference), whereas younger bulls, aged 15–25 years, invariably needed the company of one or more younger bulls to raid successfully. However, the tendency for bulls to aggregate while raiding was significant only during the finger millet season (October–December) when crops were available in plenty and cooperation, rather than competition, for obtaining this resource could be expected (Sukumar & Gadgil 1988).

In contrast, the family herds did not show any clear tendency to aggregate while raiding during any season. The only exception to this was that units of 2 elephants rarely entered cultivation since an adult female did not risk raiding with only her calf.

Table 7.4. *Group size frequencies of adult male elephants in cultivation and in the forest*

	Percentage of sightings	
Group size of adult bulls	In the forest during the day	In crop fields during the night
1	93.0	57.0
2	6.1	25.2
3	0.9	14.4
4	—	3.3
Total number of records	115	305

When these frequency distributions are treated as a 4 × 2 contingency table and tested using the G-test, the two distributions are significantly different ($G = 61.9$, d.f. $= 3, p < 0.005$).

The sightings in the forest were made over a larger area than the main study area during March 1981–February 1983; the records in crop fields are confined to the study villages during March 1981–February 1982.

7.5 Quantity consumed from crop fields

7.5.1 Rate of foraging in crop fields

In some cases the time spent by adult bull elephants in foraging on crops from a field was known. From this record the rates of feeding have been calculated (Table 7.5). When the finger millet crop was in vegetative stage and the plants were short, the foraging rate was only 1.3–1.7 kg/hour. When the fields had a high standing biomass after flowering, the quantity consumed increased substantially to over 6 kg/hour. A feeding rate of up to 12 kg/hour was recorded from a finger millet field in which the plants had low moisture.

7.5.2 Quantity and proportion of crops in the diet

The mean quantity of crops consumed by an elephant per day of raiding in two distinguishable strata of cultivation (based on differences in plant productivity and biomass per unit area, owing to rainfall, and on the intensity of feeding by elephants) is given in Table 7.6.

(a) Adult bulls

The upper limit to the weight of crops that an adult bull can consume in one night's feeding is between 70 and 75 kg of finger millet plants, recorded a number of times at Hasanur, Punjur and Kolipalya. Fig. 7.8 shows the total quantity of crops consumed by adult bulls from 10 study village enclaves each month. In the vicinity of these villages about 15 bulls were present during March–August and 20 bulls during September–February. Assuming that an average adult bull weighing 4000 kg requires 60 kg dry weight fodder (1.5% of its body weight) each day, the total quantity of crops consumed each month has been converted into a percentage of the monthly food requirement of the

Table 7.5. *Rate of feeding on crops by adult male elephants*

Crop and phenology	Total quantity consumed (kg dry weight)	Total time spent (hours)	Rate of feeding (kg/hour)
Finger millet vegetative	8.5	5.0	1.7
Finger millet vegetative	2.5	2.0	1.3
Finger millet flowering	49.2	8.0	6.2
Finger millet flowering	72.5	11.3	6.4
Finger millet flowering	56.0	8.8	6.4
Finger millet flowering/grain	72.5	6.0	12.1
Sorghum flowering	56.9	5.5	10.3

All figures refer to feeding by one adult bull.

Quantity of crops consumed

Table 7.6. *Mean quantity of crops consumed per day of raiding by an elephant*

	Mean quantity of crops consumed per elephant per day (kg)	Sample size	
		No. of days of raiding	Total no. of elephants
Adult males			
Punjur and Hasanur	43.9	40	73
All other villages	30.1	37	58
Family herds			
Punjur and Kolipalya	24.2	18	169
All other villages	10.8	15	128

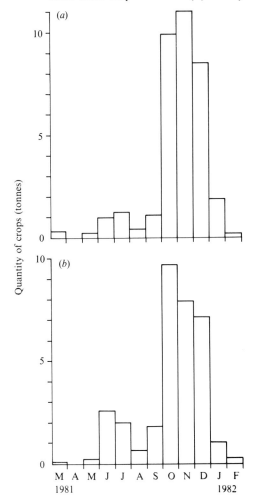

Fig. 7.8. Total quantity of crops, from ten study villages, consumed by (a) adult male elephants and (b) family herds. (From Sukumar 1989*b*.)

bull population (Fig. 7.9). The contribution of crops in the diet, which remained relatively low at 1–4% between March and September, suddenly increased to a very significant 22–30% during October–December when finger millet was cultivated. For the entire year it was estimated that cultivated crops formed 9.3% of an average adult bull's diet. Certain habitual crop raiders derived a much higher proportion of their food from cultivation. One identified bull (MA-6) foraged about 20% of its annual requirement from crop fields.

(b) Family herds

The highest quantity consumed was 52 kg per elephant per day by a herd of nine elephants from a finger millet field. Fig. 7.8 shows the total quantity of crops consumed by family herds from 10 study villages each month. It was estimated that about 200 elephants were present in the forest region around

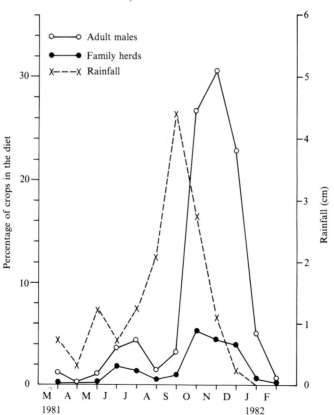

Fig. 7.9. Percentage of crops in the total diet of elephants. (From Sukumar 1989b.)

the villages during March–August and 250 elephants during September–February. An average elephant, in family herds, weighing 1575 kg would require 23.6 kg of food every day. The proportion of crops in the total diet of family herds was about 4–5% during October–December and insignificant during other months (Fig. 7.9). For the entire year cultivated crops contributed only 1.7% of the total food requirements of family herds.

7.6 Quantity of crops damaged and the economic loss

The highest area damaged in a millet field by an adult bull was 940 m^2 of finger millet crop in Hasanur. A herd of about 20 elephants devastated 16 000 m^2 of a sorghum field in Kolipalya, a mean area of 800 m^2 damaged per elephant; this was the most intensive damage by a herd recorded. In millet fields about 60% of the plants in the damaged area were consumed, the rest being just trampled or unaffected. Usually no yield was possible in trampled areas.

The mean quantity of finger millet grain lost to an adult bull and to an elephant in family herds per day of raiding is given in Table 7.7. Averages were not calculated for other millet and cereal crops such as sorghum, maize and paddy due to relatively few instances of raiding. Instead, the total quantity lost to elephant damage was assessed. Table 7.8 gives the quantity of each crop lost in the study villages between March 1981 and February 1982 and the economic value of the loss.

The total loss of millet and cereal crops included finger millet (683 quintals* of grain), sorghum (58 q), maize (32 q), paddy (35 q) and little millet (17 q);

Table 7.7. *Mean quantity of finger millet grain lost per day of raiding by an elephant*

	Mean quantity of grain damaged per elephant per day (kg)	Sample size	
		No. of days of raiding	Total no. of elephants
Adult males			
Punjur and Hasanur	60	30	59
All other villages	28	26	45
Family herds			
Punjur and Kolipalya	22	11	106
All other villages	12	16	148

* 1 q = 100 kg.

124   *Crop raiding by elephants*

65 coconut trees and 202 banana plants were destroyed. The economic value of the crops damaged in one year amounted to U.S.$18 960.

(a) Taking the total land area (both cultivated and uncultivated) of 4545 hectares for these villages, this amounted to a loss of $4.20 per hectare of land within elephant habitat. About 70% or 3200 hectares was under some form of cultivation in a year. This meant a loss of $5.90 per hectare of cultivated land.

(b) About 12 000 people (2200 households) live in the study villages. Thus a per capita loss of $1.58 was incurred (the annual per capita income in India during 1981–82 was $70 and the per capita GNP was $188 at current prices).

(c) If only the land owners are considered, there are about 900 families owning land in these villages. The loss incurred by each family averages $21. The maximum loss incurred by a farmer was $500–550 in the villages of Hasanur and Punjur.

(d) The damage caused by 15–20 adult bulls was valued at $11 760 (or $672 per bull) and that by 200–250 elephants in family herds was $7200 ($32 per elephant in a herd).

In addition to the above direct losses, the farmers would have invested at least $1000 in purchase of batteries for flashlights and fire crackers. If each cultivator or his employee spent one hour per night during 100 nights keeping a watch for elephants, this would amount to 90 000 'man hours' lost. The wage potential of this time period is $9000.

Table 7.8. *Quantity and economic value of crops damaged by elephants (March 1981–February 1982)*

Village	Finger millet Bulls (q)	Finger millet Herds (q)	Other millets and cereals (q)	Coconut (trees)	Banana (plants)	Economic value ($)
Hasanur	218	1	42	10	40	5690
Punjur	86	76	8	26	65	4175
Kolipalya	36	84	40	3	?	3210
Chikkahalli	29	24	1	2	0	1300
Neydalapuram	20	32	8	9	2	1765
Talamalai	18	52	1	15	90	2150
Mavanattam	2	2	0	0	0	80
Gaddesal	0	0	30	0	5	530
Mavallam	3	0	0	0	0	60
Total	412	271	130	65	202	18 960

Damage to millet and cereal crops refers to number of quintals of grain.
Other millets and cereals include sorghum, maize, little millet and paddy.
Other crops damaged include horse gram, sugar cane, jackfruit, mango and niger.

The proportion of the potential yield of millet crops lost due to damage by elephants varied for finger millet from 7% in Talamalai to 14% in Hasanur; it was 20% for maize in Gaddesal and 25% for sorghum in Kolipalya. In comparison to damage by elephants the depression due to variation in rainfall is usually more serious. For instance, in the villages on the western sector (Talamalai, Neydalapuram, etc.) unseasonal rains caused a depression of about 20% of the potential yield in 1981. A deficiency in rainfall may affect up to 50% of the normal crop yield. Damage by other mammals, such as wild pig, was negligible compared with that by elephant. Only in one small village, Mavanattam, did the loss to the sorghum crop due to damage by wild pigs exceed the loss due to elephant.

On the basis of the above estimates it is worthwhile to consider the total loss suffered in southern India due to elephants. Let the total elephant population be assumed as 6500 individuals (Chapter 2), of which 7% are adult bulls and the rest members of family herds. Thus, 455 bulls damage $305 760 and the other 6045 elephants in herds damage $193 440 worth of crops. The total loss in southern India would amount to about $0.5 million.

The above estimate is, of course, a simplification. It is conceivable that elephants in more fragmented habitats could cause a greater per capita damage, whereas those in relatively well-forested areas could cause much less damage. In the study area most of the damage was to millet crops, for which the loss per unit effort is much lower than for a tree crop. For instance, the maximum loss of finger millet during one night's raid by an adult bull was not more than 4 quintals of grain, worth $60. On the other hand, if 15 mature coconut trees were pushed over in one night, this would mean a loss of $500–700 in income to the farmer over the following years until the trees were replaced. If tree crops such as coconut or arecanut were the main targets in certain areas, the per capita damage by an elephant would be higher.

7.7   Methods to deter elephants, and their behavioural responses
   (a) Fire crackers: Only a few farmers can afford to use fire crackers to scare elephants. But elephants have largely learnt to recognize such psychological bluffs. On certain nights when I had the occasion to witness fire crackers being used against crop raiders, I noticed that the elephants ignored these displays and continued their feeding. Morris (1958) describes how 'bamboo gun rockets' were successfully deployed by him to chase elephants from his crop fields in the Biligirirangans. This device is not used at present in the region.
   (b) Shooting: Firing with a gun over the elephant may be more

effective, but here again the response may not always be the same. I have seen raiding elephants ignore even gun shots. Only when directly fired upon did they leave the field.

(c) Loudspeakers: In one farm a tape-recorded jumble of noises played through a loudspeaker was effective in keeping away a raiding bull elephant.

(d) Vehicles: A tractor or a jeep fitted with spotlights is fairly effective in dislodging elephants provided the vehicle can be taken close enough to the animal. But once the vehicle is withdrawn the elephant may come back.

(e) Trenches: Trenching has largely proved ineffective in keeping away elephants. Well-designed trenches are rarely dug. Elephants may fill up a trench by digging the soil with their forefeet. The few trenches inspected were all deficient and elephants had crossed every one of them (Fig. 7.10). Improperly maintained trenches quickly get filled and are rendered ineffective.

(f) Electric fences: Occasionally farmers run a strand of wire along the periphery of the field and illegally electrify it from the 230 V mains. Apart from the danger to people, animals coming into contact with the wire usually suffer a fatal shock. A few elephants die each year in this manner. The high-voltage electric fence of the non-fatal type was first used in a coffee estate in the Biligirirangans many decades back, but this was later discontinued. This fence has been revived in recent years in southern India. The fence design and its effectiveness are discussed in more detail in Chapter 12.

Most farmers do not attempt anything more than merely shouting and shining flashlights. Platforms built on tree-tops are usually used by them to keep guard at night (Fig. 7.11). Sometimes, flimsy thatched structures are built on the ground. People keeping watch from here are highly vulnerable to attack by elephants.

7.8 Crop raiding in rain forest habitats

The account by Blair *et al.* (1979) about elephant depredation in the Federal Land Development Authority (FELDA) schemes in peninsular Malaysia is the main source of information. The main targets of damage by elephants are oil palm (*Elaeis guineensis*) and, to a lesser extent, rubber (*Hevea brasiliensis*). Although a tendency exists for attacks to increase during the northeast monsoon in the later months of the year, the relationship

Fig. 7.10. A trench which, after completion, was crossed by a bull elephant.

is not consistent (Fig. 7.12). The series of sharp peaks and troughs in the frequency of damage, accompanied by shifts in attack from one FELDA scheme to another, may be due to many reasons. A few elephants may be responsible for all the damage. These elephants may not show any definite pattern of movement in the relatively aseasonal, homogeneous rain forest habitats. Unlike the seasonal cultivation of millet and cereal crops, plantation crops are perennially available. Hence, the frequency of attack on plantations may peak during any time of the year as and when an elephant herd moves near a scheme area. The shift in attack from Simpang Wa Ha to Bukit Easter and later to Lok Heng Timur, planted in 1975, 1976 and 1977 respectively, is apparently a transfer from older to younger schemes.

Elephants show a distinct preference for attacking oil palms of 1–2 years (Fig. 7.13). The average age of damage was 1 year 10 months. There was little damage after the fifth year. Damage by other mammalian pests such as wild pig and porcupine may also be considerable. A total of 2.9 million oil palm and rubber trees had been damaged by these three mammals until 1978. Elephants were responsible for 58.6%, wild pig for 30.7% and porcupine for

Fig. 7.11. Tree-top platform used to keep guard over cultivated fields. People who stay in thatched huts at ground level are in danger of being killed by elephants.

Crop raiding in rain forests 129

Fig. 7.12. Frequency of damage to FELDA oil palm and rubber plantations in peninsular Malaysia. Source: Blair *et al.* (1979).

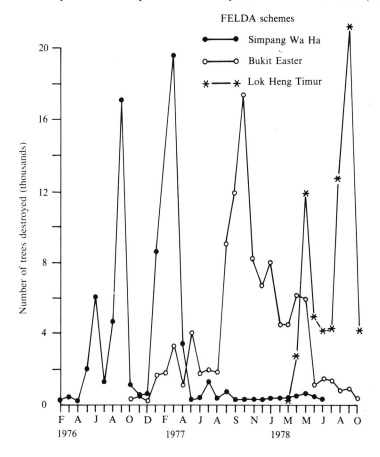

Fig. 7.13. Age of oil palm plants at time of attack by elephants. Source: Blair *et al.* (1979).

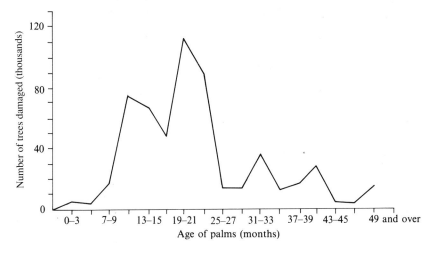

Table 7.9. *Cost implications of damage to oil palm plantations*

Items of cost	Cost/hectare ($) by age of palms at time of damage (months)			
	3	9	21	60
Establishment costs (E) including planting material and labour	173	173	173	173
Operational costs (O) including weed control and manuring	38	115	348	1174
Administrative costs (A)	32	41	59	118
Total EOA costs	243	329	580	1465
Settler income lost	31	92	1551	3405
Duty lost	17	52	797	1717
Export value lost	62	185	2819	6074

An exchange rate of 2.3 Malaysian dollars per US dollar has been used in the conversion.
Modified from Blair *et al.* (1979).

10.7% of the damage. The age of trees damaged by wild pig and porcupine is, on average, lower than that in cases of elephant attack. Hence, the economic loss due to elephant is proportionately higher.

The cost implications of destruction to oil palm plantations of different ages is given in Table 7.9. The loss incurred per elephant can be extremely high. In the Johore Tenggara area a herd of nine elephants is supposed to have destroyed a total of 4400 ha of oil palm and 180 ha of rubber plantations within six years, resulting in a loss of about U.S.$37 million. Although there is no proof that a mere nine elephants were responsible for all this damage, it is clear that the 'costs' of maintaining an elephant in the wild can be unacceptably high. About 4000 ha of FELDA's plantations, valued at $20 million, are destroyed annually by elephants. This represents 10% of the new land area developed by FELDA each year.

7.9 Causes of crop raiding

Raiding of agricultural fields by elephants can be explained in terms of proximate factors such as contact with cultivation, especially in fragmented habitats, in the course of their movement for foraging or drinking. However, in ultimate terms crop raiding can be thought of as an extension of its natural optimal foraging strategy.

7.9.1 Raiding in relation to movement patterns

In the first instance, elephants probably come into contact with cultivated land in the course of their natural seasonal movement. There is evidence that the movement pattern seen today in the study area is as close to a natural situation as possible in spite of habitat changes due to human impact. From the observations of Sanderson (1878), more than a hundred years ago, it can be seen that at least one elephant clan was following the same strategy as that seen in recent years. Elephants used to move from the moist deciduous forests of Zone 3 in September into lower-elevation habitats, such as those around Punjur and Kolipalya, even when the former village was very small and the latter did not exist (discussed in Chapter 4).

A comparison of the frequency of raiding in each village with the overall strategy of movement and the density of elephants in the zones adjoining these villages shows an obvious correlation. This can be illustrated with a few examples. During September, when elephants began moving southwards from Zone 3 into Zone 7, raiding began first in the villages of Punjur and Kolipalya. Some herds also moved further south into Zones 8, 9 and 10. There was sequential raiding at Chikkahalli, Neydalapuram, Talamalai and Mavanattam corresponding with the movement of herds into adjacent zones during October–November. Hasanur was free of raiding by herds until the end of November, although the finger millet crop was in a highly preferred stage during October–November, because no herds were present in its vicinity. Similarly, in the southern plains (Zone 18) raiding in villages such as Pudukuyyanur, adjoining the forest boundary, was confined to between late November and February, in spite of crops being grown almost all year round owing to better irrigation. This correlates with the presence of elephants in the adjacent forests from only November to February.

In contrast to family herds, many adult bulls seem to be habitual crop raiders. Certain bulls took up position for long periods near villages, regularly raiding the fields at night and retreating into the forest during the day. For instance, one identified bull (MA-6) spent the entire period from June to early January near Hasanur, raiding whatever crops were available, after which it headed west towards Talamalai where it continued its depredation.

7.9.2 Competition for water

Village ponds and small irrigation reservoirs are often used by elephants at night. Because the region is relatively scarce in water, a need to utilize water from agricultural land increases the frequency of an elephant's contact with crops. In the process the traversed fields may be damaged. Elephants seem able to smell water and move towards a large water body or an area receiving rain (Leuthold 1977; Allaway 1979). Herds of elephants which traversed the fields of Gumtapuram and Kolipalya often headed towards the Suvarnavati Reservoir, the largest water body in the area. Small ponds in the villages of Gaddesal, Hasanur, Talamalai and Chikkahalli were used by bulls and herds at night. Visits to ponds were often combined with attempts to feed on available crops. Whether elephants damage crops in their passage to water or whether the utilization of ponds is merely opportunistic in the course of raiding crops cannot be deduced with certainty. Both these situations may be true, depending on the specific requirements and motivational status of the animals concerned.

7.9.3 Reduction and fragmentation of natural habitat

Whereas a large and compact natural habitat allows unimpeded movement, with cultivation making inroads there is often little room left for manoeuvrability by these far-ranging mammals. The larger the size of a cultivated enclave, the longer its perimeter abutting the forest and the higher the frequency of raiding that can be expected, owing to the higher probability of an elephant making contact with its boundary. As expected the frequency of raiding was, in general, directly proportional to the size of the cultivated enclave. The situation may be aggravated when their passage from one habitat to another becomes narrow. For instance, the traditional movement has been hampered by the settlements of Kolipalya and Punjur. Since the steep hills to the east of Punjur are not much used, most of the elephants are forced to move through the narrow (1.5 km) corridor between these two enclaves. This funnelling effect increases their contact with cultivation.

The acute problem of crop raiding in the Bannerghatta–Anekal sector of the Eastern Ghats can be certainly attributed to the drastic reduction in habitat. The elephant habitat has been reduced to a long, narrow, convoluted belt of scrub only 1–2 km wide in places. Given their high mobility, elephants cannot but encounter cultivation once they move through this stretch. If chased away from one side they can reappear within a short while on the other side.

Similarly, elephants confined to a small isolated habitat can be expected to indulge frequently in raiding crops. Fragmented patches provide a convenient base for crop raiders. At least two crop-raiding bulls often sought refuge in the Akkurjorai Reserve, which is isolated from the main forest area (see Fig. 7.1). There was also the curious case of a bull group in the Kattepura Reserve in Karnataka. This forest patch of 234 ha is surrounded by coffee plantations and a reservoir. The elephants here were in constant conflict with the human settlements. Eight elephants were captured in March 1987 by

Causes of crop raiding

chemical immobilization. All of them were male elephants (one was a sub-adult). These elephants had strayed in from nearby forests.

7.9.4 Degradation of habitat

Degradation of the natural habitat is often mentioned as the primary cause of crop raiding. Here, the term degradation is used to describe any exploitation of the habitat that reduces the elephant's food resources. A qualitative ranking of habitat types around the villages, when compared with the intensity of raiding, showed no clear pattern. Elephants inhabiting an area with a clear surplus of natural food resources would still resort to crop raiding. Even if there were just one bull in a large expanse of forest it could be expected to raid crops to some extent. Although a degraded area in the vicinity of a village might provide motivation for entering cultivation, the tendency for elephants to avoid below-optimum habitats might be a counter-force, taking them into more frequent contact with cultivated tracts situated in the more favoured natural habitats.

7.9.5 Palatability and nutritive value of crops

Ultimately, if a crop-raiding strategy is to confer some benefit on the elephant, it must provide a higher level of nutrition than foraging on wild plants. Because humans have selected their food crops primarily on considerations of sensory quality, digestibility, absence of toxins, productivity and, in recent times, for their nutritive value, it is not surprising that many such crops are also attractive to elephants, especially if these crops are analogous to their wild counterparts.

The elephant's natural preference for plants of the Gramineae, Palmae and Leguminosae could also be extended to the analogous cultivated plants. When it encounters a cultivated grass field, the elephant treats it as any wild grass species. Elephants are known to select only the central rachis while feeding on wild palms (Olivier 1978*b*). In the cultivated coconut palm and oil palm also the central rachis is consumed. Similar selection for wild and cultivated Leguminosae, Anacardiaceae and Moraceae occurs.

Cultivated crops, as a whole, have higher palatability and nutritive value than their wild counterparts (Appendix III). During the second wet season the wild grasses *Themeda* and *Cymbopogon* are very fibrous and siliceous, whereas the succulent finger millet plants are much more edible. Sucrose in sugar cane could appeal to the elephant's palate. Cultivated grasses are also either low in or devoid of secondary compounds. These may be more appealing than the terpenoid-oil-containing *Cymbopogon flexuosus*, which is a common wild grass in the area.

The crude protein content of finger millet in inflorescence stage (8.3%) and grain stage (5.3%), or of paddy (10%), is not very different from that in the leaves of tall grasses (6.8–8.0%) during the second wet season. However, at this time elephants do not consume the leaves but the basal portion of wild grasses, which has only 2.0–3.8% protein. Hence, feeding on cultivated grasses provides them with substantially more protein.

The calcium content is also higher in finger millet inflorescence (10.8 mg/g) and grain stage (7.4 mg/g) than in the basal portion of wild grasses (0.8–2.3 mg/g). The sodium content of finger millet in inflorescence stage (0.94 mg/g) is strikingly higher than of any other wild food plant, including barks, analysed (range 0.10–0.28 mg/g). Paddy in the mature stage also has a relatively high sodium content of 0.36 mg/g. As mentioned in Chapter 5, this could be important to elephants, which are likely to be deficient in sodium.

Elephants may respond to the nutrients associated with the phenological stage of a cultivated grass. From the time of pollination, which usually occurs two or three days after flowering, there is a sudden influx of sucrose and amino acids from the leaf into the developing seed for about 15 days. After this the grain begins to dry. It may be significant that peak raiding of finger millet fields occurs during the inflorescence stage.

The higher propensity of adult male elephants than of female herds to raid crops may have its origin in the higher variance in male reproductive success in this polygynous mammal, leading to selection pressures favouring a risky strategy in the males to derive better nutrition for enhancing reproductive success (Sukumar & Gadgil 1988). Extra nutrition from crops could contribute to better growth, adult body size and a successful expression of musth, all of which may mean increased dominance over other bulls and better access to females for mating. I do not argue that the additional risks (injury or death during raids) taken by male elephants in crop raiding are necessarily offset by any higher reproductive success derived thereby, but merely that the higher level of raiding by males is a consequence of their 'high risk–high gain' strategy, moulded by natural selection, to enhance reproductive success.

8

Manslaughter by elephants

One of the tragedies of the interaction between elephants and people is that elephants kill people even without any direct provocation. In India the number of people who die of rabies after being bitten by dogs is far greater than those who fall victim to elephants; about 15 000 people die of rabies annually compared with 100–150 people who are killed by elephants. Yet public outcry has been and will continue to be more vociferous against 'rogue' elephants (or man-eating tigers) than against stray dogs.

What are the circumstances under which people are killed by elephants? Are some elephants more prone to kill than others? Details of more than 150 cases of manslaughter by elephants in southern India provided answers to these questions.

8.1 **The people killed**

Between 30 and 50 people are killed by elephants every year in southern India. Only half the cases are officially registered. Records maintained in some other Indian states reveal that on average about 5 people are killed in Uttar Pradesh and 30–40 in Assam. During the 1970s about 10 people were killed annually in West Bengal but during 1986–87 this had increased to 35–40 per year (Lahiri Choudhury 1980 and personal communication). These figures indicate that between 100 and 150 people fall victim to elephants annually in the country. No figures are available for other countries, although a newspaper report in April 1987 mentioned that wild elephants raiding villages in Bangladesh killed 9 people and destroyed 30 houses.

Records obtained from southern India (Table 8.1) showed that a majority of the people killed were adult men (77%), with relatively few victims being adult women and children. This is simply due to the higher frequency of

Table 8.1. *Patterns in manslaughter by elephants*

	Number	Percentage
The people killed		
Men	102	77.3
Women	23	17.4
Children	7	5.3
Total	132	
Place of encounter		
Forest	68	55.3
Cultivation	55	44.7
Total	123	
The elephants responsible		
Male	51	82.2
Female	6	9.7
Herds	5	8.0
Total	62	

contact between elephants and men. Most of the victims hailed from villages within or near elephant habitat and included farmers, graziers and labourers.

8.2   The elephants responsible

It is often impossible to determine the sex of the elephant responsible without any doubt because this knowledge may have been restricted to the victim. Sometimes the sex may be indicated by circumstantial evidence; for example, an extremely large footprint invariably points to an adult bull, a gored injury to a male tusker and footprints of various sizes to the involvement of a herd, though not necessarily a female elephant.

The only instances where the sex of the elephant is clearly known is when people accompanying the victim witness the encounter. Out of 62 cases where this was determined, an extremely significant proportion (82%) involved sub-adult or adult male elephants. As there are fewer sub-adult and adult males than females in the population (Chapter 11), it is clear that male elephants are far more prone to aggressive behaviour culminating in manslaughter. Even among bulls there is wide individual variation in aggressive tendencies. It is certainly true that only some bulls are habitual killers. In the study area there were two identified 'rogues', one of which had killed numerous people within cultivation. Similarly, during December 1982 and January 1983, an adult tusker killed five people in separate incidents near the town of Gudalur in Tamilnadu State.

8.3 Circumstances of encounter

Encounters between elephants and people take place in the forest or within human settlements. Out of 123 cases in which the place was known, 68 cases (55%) occurred in the forest and a significant 55 incidents (45%) took place within settlements.

Incidents in the forest usually involved wood-gatherers (often women) or cattle graziers. People walking along roads in the forest, such as labourers going to work, pilgrims visiting a temple within the forest or simply people going to a village, are also at risk. There have been instances of photographers getting killed when approaching an elephant closely, mentally deranged persons walking up to an elephant oblivious of the danger, a deaf person caught unawares and even a drunken man trying to hold a bull by its tusks.

Because the undergrowth in a forest may be dense, people often tend to walk along paths which have been created by the constant use of elephants. Owing to poor visibility, which may be only 5–10 m, one is likely to encounter elephants at close range. If an elephant's immediate reaction is to charge there is neither time nor place for safety. Most of the killings within the forest took place during the daytime.

By contrast, encounters within settlements were almost invariably at night. Male elephants were involved in 23 out of the 25 such instances in southern India where the sex was determined. In other regions, such as West Bengal, elephant herds have been reported to have killed people by demolishing their huts (Lahiri Choudhury 1980). Certain bulls respond aggressively to any attempt to chase them away from crop fields. A flashlight shone at them often evokes a charge. This may be followed by an attempt to pull down the hut from where the light was flashed (Fig. 8.1). People sitting on tree-top platforms are safe, but those who foolishly guard their fields from flimsy thatched structures at ground level are highly vulnerable.

Elephants may also react aggressively to a dog's bark. Farmers sometimes use dogs to warn them of the presence of raiding elephants. If an elephant chased the dog, the latter would naturally tend to run back to its master, bringing behind it an elephant which might redirect its aggression on the person.

Because there are few reliable witnesses to the actual method of killing, this must often be construed from indirect evidence. An elephant may use its trunk, foot or tusk to kill a person. At the end of a charge an elephant may lash out its trunk (Fig. 8.2). It may also grasp the person with its trunk and fling him away with considerable force. Most of the victims seem to be killed owing to such actions (Fig. 8.3). An elephant may kick with its front foot or trample the victim deliberately or inadvertently, especially if the person falls

Fig. 8.1. A house damaged by a bull elephant. The woman and her child, who were inside at the time, escaped injury.

Fig. 8.2. A charging female elephant.

Circumstances of encounter 139

Fig. 8.3. Jungu of Hasanur was killed inside the forest by an elephant. A tuskless elephant grasped him with its trunk and flung him to the ground. There were no apparent external injuries to his body.

Fig. 8.4. A bull elephant in musth. Note the secretion from the temporal gland between the eye and the ear.

down while being chased. Male elephants may sometimes gore the victims with their tusks. Such bizarre behaviour as tearing the body into pieces or grinding the remains with the foot is very rare, notwithstanding the gory tales that people commonly believe.

8.4 Causes of aggression

Aggressive behaviour in an animal may be governed by both genetic determinants and environmental influences. Lorenz (1966) proposed a 'dynamic instinct' view of aggression. An aggressive trait is basically inherited and could be explained in terms of Darwinian fitness. Genetic determinants of aggression can function through the endocrine and neural processes translated into the precise muscular movements of aggression in response to an appropriate stimulus.

It is well known that a rise in the level of the male sex hormone, testosterone, is accompanied by increased aggressiveness in many vertebrate groups (see Leshner (1978) for a review). As a first approximation it can be expected that male elephants will be innately more aggressive than female elephants. Among Asian elephants this is true of intra-specific aggression (male–male competition) and inter-specific aggression towards people, resulting in manslaughter. Male elephants are also more aggressive during *musth* (Fig. 8.4), when their physiology and behaviour show significant changes (Eisenberg, McKay & Jainudeen 1971; Jainudeen, McKay & Eisenberg 1972). The level of testosterone in blood plasma increases from 0.2–1.4 ng/ml during the non-musth phase to 29.6–65.4 ng/ml during the full musth phase in captive Asian bull elephants (Jainudeen, Katongole & Short 1972). Captive bulls in musth are extremely difficult to handle even by experienced trainers (Williams 1950). Bulls in musth show increased aggression towards other bulls (Kurt 1974). There is, however, only anecdotal evidence that a captive bull may be more likely to kill people when it is in musth.

Experimental psychologists have stressed the role of experiences in the development of behaviour. Aggressive behaviour is learnt gradually when an animal undergoes pain during competition for resources or during play-fighting (Scott 1958, 1962). It has been well documented that animals respond aggressively to pain (Ulrich 1966). This principle may apply to aggression by elephants. A broken tusk may result in a persistent 'toothache'; such elephants may be prone to increased aggression. A bull (MA-6) whose right tusk was broken near the lip line was one of the most aggressive bulls in the study area (Fig. 8.5). Elephants injured by bullets or in encounters with man have been implicated in manslaughter (Krishnan 1972; Caras 1975). Thus, elephants that have survived after being shot during crop raiding or by hunters may show aggressive tendencies. Frustration or interruption of a goal-oriented behaviour may also provoke an aggressive reaction (Dollard *et al.* 1939). A large number of killings occur within settlements when people try to frustrate the elephant's goal of feeding on cultivated crops.

The elephant's experience with people has varied widely through space and time from peaceful coexistence to brutal slaughter. The intensity of the elephant's aggressive response to man could be expected to reflect the nature of this interaction. In regions where elephants have been allowed to live in peace they can be expected to be less aggressive towards people, as compared to places where they have been continually harassed by hunters and by encroachment on their habitat. General observations on the African elephant seem to verify this expectation. Elephants are not aggressive towards people in the Amboseli National Park, Kenya, which the Masai people and elephants share without conflict (C. Moss, personal communication). By contrast, adult female elephants were often very aggressive in Lake Manyara National Park, Tanzania, where poaching was common (Douglas-Hamilton &

Fig. 8.5. A bull elephant (MA-6) with its right tusk broken near the lip line. This elephant was an aggressive crop raider.

Douglas-Hamilton 1975). It must be emphasized, however, that elephants belonging to a single population and region show wide differences in behaviour ranging from timid to aggressive.

From an evolutionary view-point, inter-specific aggression in elephants could have largely arisen as an anti-predatory strategy. Extinct species of Proboscidea, such as the mastodon and the mammoth, were potential prey for contemporary carnivores including the sabre-tooth tiger, the scimitar-tooth cat and the sabre-tooth cat. Even today, young elephants are in danger of predation by lions in Africa and tigers in Asia. Natural selection would have favoured group defence and aggressive threat displays.

Man has also been a predator on the Proboscidea since the Pleistocene. The mastodon and the mammoth were hunted by prehistoric people. This predatory role continued with the regular capture of *Elephas* in Asia, beginning at least 4000 years ago, and the hunting of *Loxodonta* in Africa. On an evolutionary time scale the role of people in shaping elephant behaviour through selection may not be significant. Therefore, aggression by elephants towards people may be primarily an extension of their natural anti-predatory behaviour. Although an anti-predatory strategy would be true of the matriarchal family herds, it would not be necessary for adult bulls. Aggressive behaviour by adult bulls resulting in manslaughter could, however, be explained by other considerations. A basic aggressive trait could be maintained in the males by genetic mechanisms and by cultural transmission during the pre-pubertal years when they acquire experience within the family. In addition, the higher propensity of male elephants than of females to raid crops leads to a more intensive interaction with people within cultivated land. Male elephants that are aggressive crop raiders may also exhibit similar behaviour towards people in the forest.

9

Habitat manipulation by people

The most serious impact of human activity on the elephant's habitat has been the reduction and fragmentation of the habitat, resulting in the compression and isolation of elephant populations. This has been accompanied by exploitation of the habitat for various human needs such as timber, fuel wood, food, water and grazing area for livestock. On a historical time scale, this impact has certainly been detrimental to the elephant, as it has been to the overall biological diversity. Viewed from a restricted angle, however, not all human activities have necessarily been incompatible with the elephant's use of the habitat. In local areas, people may even have served to improve, albeit temporarily, the quality of the habitat for elephants and other herbivores. It is generally accepted that secondary plant communities often support a higher biomass of certain herbivorous mammals than do primary climax communities.

Is conversion of primary rain forest to secondary vegetation desirable? How does exploitation of the habitat affect the elephant's requirements, either negatively or positively? Can human needs be met without detriment to the elephant? These are some of the questions discussed in this chapter in the context of human activities such as shifting cultivation, timber extraction, plantation forestry, cattle grazing and dam construction. Only the broad principles are presented, with examples mainly from southern India. Available data from other elephant regions are also presented. Obviously each region has to be evaluated separately for local management policies to be formulated.

9.1 Reduction in habitat area and fragmentation

The process of reduction in habitat area, operating over a historical period and continuing at present, has been described in Chapters 1 and 2.

Between 1880 and 1980 about 15% of the habitat in the study area was lost to agriculture and reservoirs. Excluding the sub-optimal montane *shola* forests and grasslands (above 1500 m), the elephant's range in southern India has shrunk by perhaps 20–25% within a century. Reduction in habitat compresses the existing elephants into a smaller area at higher densities. Compression has been implicated in a high rate of damage to trees by elephants and the conversion of woodland to grassland (Chapter 6). In many regions of Asia the increase in elephant densities due to compression of populations may have been partly offset by capture of elephants or their elimination. Although damage to habitat may not be a major issue, the reduction in population numbers means a reduction in viability of the population (see Chapter 12). Fragmentation of the habitat has also brought the elephant into increased contact and conflict with people. Elephant bulls or herds isolated in fragmented habitat patches invariably resort to raiding cultivated crops (Chapter 7).

9.2 Shifting cultivation

Shifting cultivation has been a way of life in hill forests all over tropical South and Southeast Asia. The impact of this practice on the habitat has varied from one region to another. In southern India many tribes have practised shifting cultivation for centuries. The Sholaga tribe cultivated sites within the moist deciduous forests in the main study area until the 1960s.

Sites of cultivation are generally rotated after 2–3 years of use, with a rotation period of 5–20 years or more, although no definitive information is available for the study area. Abandoned sites show successional stages commencing with herbs and shrubs. A mosaic of such sites with different sucessional stages may be seen. If there is no further interference the ecological succession may proceed to a climax vegetation; more often, owing to fire and other factors, this may be only a sub-climax (Olivier 1978*b*). Such areas attract herbivores, including elephants, which feed on grass and young trees. In the study area, abandoned sites of cultivation which were inspected had regenerated into forest. No attempt was made to quantify the availability of food plants in these areas compared with natural forest, because the surrounding forests had been logged and could not serve as controls for the comparison. In southern Indian forests there is, however, the danger of pernicious weeds such as *Lantana camara* and *Chromolaena odorata* invading cleared areas and replacing favoured plants. Shifting cultivation practised on a small scale may not damage the habitat for elephants.

The story has been quite different in central and northeast India. Extensive slash-and-burn shifting cultivation with short rotation periods, often less than

5 years, has degraded the forest cover over large tracts of elephant habitat (Lahiri Choudhury 1980). Nutrient levels under a 5-year cycle are significantly lower than under 10–15 year cycles. Short cycles also result in rapid soil erosion and desertification of sites in high-rainfall areas of northeast India (Fig. 9.1). Plant succession is arrested, with weeds such as *Chromolaena adenophora* and *Imperata cylindrica* dominating (Mishra & Ramakrishnan 1983*a,b,c*). Extensive areas of grassland without effective tree cover would be a sub-optimal habitat for elephant (Chapters 5 and 6).

9.3 Extraction of plant products
9.3.1 *Timber felling*

The pattern of exploitation of forests for timber, pulp wood, fuel wood and other categories varies with the vegetation type in southern India.

The moist forests are valuable sources of timber and soft woods for industry. Species extracted from evergreen forests include hard woods such as *Mesua ferrea, Hopea parviflora, Cullenia excelsa, Palaquium ellipticum, Calophyllum elatum* and *Vitex altissima*, and soft woods such as *Persea macrantha, Bombax malabaricum, Vateria* sp., *Polyalthia fragrans, Holigarna* spp., *Myristica malabarica* and *Ficus* spp.

Fig. 9.1. Shifting cultivation in elephant habitat in northeastern India. The lack of tree cover over vast areas has rendered these habitats sub-optimal for elephants. (Photo: D. K. Lahiri Choudhury.)

The moist deciduous forests are exploited for important timber trees such as *Dalbergia latifolia* (rosewood), *Tectona grandis* (teak), *Terminalia tomentosa, Pterocarpus marsupium* and *Grewia tiliaefolia*. There seem to have been extensive unorganized fellings in most regions of the erstwhile Madras Presidency from the mid-nineteenth century until the early twentieth century. For instance, Joseph (1969) records a 'disastrous felling' of 30 000 cubic feet of rosewood from the Minchikuli valley (Zone 14) and Gaddesal (Zone 15) during 1906–7. Fellings in Mudumalai were similarly unregulated until 1910 (Hicks 1928). Between 1860 and 1884 about 600 000 cubic feet of *Tectona grandis, Dalbergia latifolia* and *Pterocarpus marsupium* from Mudumalai had been sold and a large unknown quantity wasted. It was only after 1910 or so that a system of compartments was introduced. Each compartment was to be exploited once in 30 years. This system is followed in most regions today.

Selective logging in a closed canopy forest alters the micro-climate, soil properties and vegetation structure. Pioneer and light-demanding species, especially *r*-selected plants, establish in the disturbed area. Some of these invading plants, such as grasses and bamboos, are useful food plants for elephants. If logging is done according to sound silvicultural practice (which is often not so), the resulting habitat changes may even favour elephants. Selective felling in the study area does not seem to have caused any adverse habitat change for elephants. The same cannot be said of clear felling for raising monoculture plantations. Even selective felling over extensive areas of primary forest may not be desirable from other points of view, as discussed later.

The dry forests have been exploited extensively for fuel wood and various categories for 'small timber'. One of the first and most valuable trees to be extracted from the study area by the Forest Department was the fragrant *Santalum album* (sandalwood), characteristically found in the dry tracts. Sandalwood is the main source of revenue to the Forest Department in Tamilnadu. Poaching of sandalwood is common. Sandal extraction does not affect the elephant as it is neither a food plant nor does its removal cause any adverse habitat change.

During the nineteenth century, the Forest Department of the Madras Presidency allowed fuel wood to be taken free of cost. All species except for teak, rosewood, sandalwood and tamarind were allowed to be cut. Fuel Working Circles in the Satyamanglam Division, established in 1908 in the plains and later extended to the plateau, had a varied history with 'diametrically opposite prescriptions from time to time' (Joseph 1969). During the Second World War the Reserved Forests of the Madras and

Bombay Presidencies were opened for charcoal manufacture. Extraction continued on a large scale in Satyamangalam until Joseph (1969), commenting on the extremely poor stand, prescribed reducing the area to 7550 ha or 5.6% of the area. The next Working Plan by Kala (1979) reduced this further to 1632 ha. Removal of fuel wood by the cartload in the forests of Karnataka was permitted until the 1960s. As an example, the quantity of wood removed from the composite Mysore Division (which included the Chamarajanagar Division) in one year (1962–63) was 44 120 tonnes of dry wood and 15 370 tonnes of fresh wood.

Added to the department working, the removal of wood by local villagers for their domestic use and for sale has further altered the natural vegetation. The areas which have been degraded into semi-scrub in the study area include Zones 1, 7, 9, 10 and 18. The standing biomass of vegetation in reserve forest areas adjoining cultivation is visibly lower than in interior areas. This would reduce food availability for elephants in the peripheral dry thorn forest tracts.

9.3.2 Bamboo extraction

The two bamboos in the study area are *Bambusa arundinacea* and *Dendrocalamus strictus*, the former confined to stream banks and moist localities and the latter occurring in the drier regions. Bamboos are perennial, woody grasses showing exponential growth and monocarpic flowering. *B. arundinacea* has a flowering cycle of 45–50 years; *D. strictus* has a cycle of around 32 years and flowers sporadically (Prasad & Gadgil 1981; Gadgil & Prasad 1984). After a massive seed production the plant dies.

The entire cohort of *B. arundinacea* in a region flowers gregariously; thus the standing biomass in a locality follows a cyclic pattern with a perid of 45–50 years. This has important implications for the food supply of dependent herbivores. In the period immediately following gregarious flowering in a region, the standing green biomass would be extremely low until the new seedlings attain sufficient growth, which might take 5–10 years. On the other hand, since the entire population of *D. strictus* in an area does not flower at the same time, there would always be sufficient green biomass in vegetative stage. In the study area, *B. arundinacea* flowered gregariously in 1927–28 and again in 1972–74.

Prior to 1970, the Bamboo Working Circle in the Satyamangalam Division covered 50 000 ha or 37% of its total area. The Working Plan for 1970–80 (Joseph 1969) envisaged an annual supply of 8500 tonnes from the Division for the paper industry. The quantities of bamboo removed from the Satyamangalam and adjacent Erode Forest Divisions between 1971–72 and 1981–82 are shown in Fig. 9.2. The drastic reduction in harvest is partly a reflection of the depletion of bamboo in these regions. For instance, the effective area under bamboo during a 10-year period between the Working Plans of Joseph (1969) and Kala (1979) came down from an estimated 630 to 180 km², a decrease of 71% from the original area. Since *Bambusa* flowered during this period the reduction could be partly explained by its death which follows the flowering phase. But *Bambusa* accounted for only about 5% of the effective area in 1969. Hence the decline was mostly in the area under *Dendrocalamus*.

Bamboo resources have been largely overexploited in southern India (Prasad & Gadgil 1981). Contractors flout most silvicultural norms while extracting bamboo. Entire clumps are cut with the rhizome and sometimes the area set on fire to hide the destructive practice. Green and immature culms are cut along with dry, mature ones. Although feeding by elephants may partly account for the poor regeneration, human factors including fire, cattle grazing and overexploitation are mainly responsible for the present state of affairs. There is also the possibility that noxious weeds such as *Lantana* and *Chromolaena,* which have gained a firm hold in the area, may be occupying the niche once held by bamboo and suppressing its regeneration.

In evaluating the impact of bamboo exploitation on the elephant, its

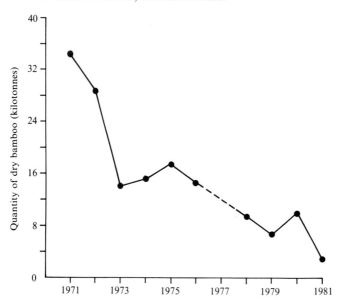

Fig. 9.2. Levels of bamboo harvest in the Satyamangalam and Erode Forest Divisions, southern India.

feeding behaviour must be considered. Although elephants do feed on the bamboo culms, they show a preference for the young regenerating branches and leaves. A high standing biomass of bamboo culms consisting mainly of tall mature culms does not necessarily meet the elephant's requirement in the best manner. A younger crop of regenerating bamboo with a profusion of young culms, culm branches and leaves is of better food value. Elephants themselves tend to keep bamboo clumps in this growth form through constant browsing.

9.3.3   *Grass extraction*

People all over the tropics have traditionally exploited grasses from the natural habitat for use as fodder for their cattle or material in constructing houses. The levels of harvest are usually negligible in relation to the productivity.

The dominant perennial grasses in the study area are species of *Cymbopogon* and *Themeda*. These are in demand by the paper industry for pulp manufacture. Between 400 and 800 tonnes of air-dried grass were annually harvested from Satyamangalam Division and about 2000 tonnes from Chamarajanagar Division during 1981–83. Grass was extracted along roadsides from Zones 3, 11 and 16. The harvest amounted to between 5% and 15% of the net primary production of grass in these zones.

Removal of grass has both positive and negative impacts. Grass is usually cut only during the dry months between January and March. This reduces the incidence of fire, thus fulfilling a management objective of the Forest Department. On the other hand, the cutting of grass may destroy saplings in the undergrowth, allow weeds to invade and reduce food supply in zones which elephants are likely to occupy during the dry season. The poor growth of grasses in certain habitats such as Zone 11 may not warrant further disturbance. Further research is needed before firm recommendations can be made on this issue.

9.3.4   *Minor forest produce collection*

A variety of 'minor forest produce' is exploited from the study area. These include tamarind pods (*Tamarindus indica*), soap-nut (*Sapindus emarginatus*), gall-nut (*Terminalia chebula*), gooseberry (*Phyllanthus emblica*), wood-apple (*Limonia acidissima*), mango (*Mangifera indica*), neem fruit (*Azadirachta indica*), pods of *Acacia sinuata*, bark of *Acacia* spp., *Cassia fistula* and *C. auriculata*, fruits of *Ziziphus* spp., leaves of *Phoenix humilis*, lichens and mosses. The collection and sale of these products are controlled by the Forest Department through local tribal cooperative societies.

Collection of many of these products does not affect the elephant's food supply. Removal of *Acacia* bark will have to be regulated, as the *Acacia* shrubs are important in the elephant's diet during the dry season. Leaves of *Phoenix* are also eaten by elephants but the quantities harvested were negligible. Fruits of tamarind, wood-apple and mango are consumed seasonally. These are not major components in the diet but it would be prudent to leave a certain proportion uncollected for consumption by elephants.

9.4 Grazing by domestic livestock

Grazing by livestock is widespread in the tropics. The numbers of cows, buffaloes and sheep permitted to graze in the Satyamangalam Division during 1982–83 added up to an average density of 28 domestic animals/km^2. Because unlicensed livestock are also taken in for grazing, the actual numbers may be up to 50% higher. Ecological densities in certain zones were 50 livestock/km^2 or higher.

The impact of intensive livestock grazing on the habitat is well known and will not be elaborated here. The soil is compacted by trampling; infiltration of water into the ground is low and surface runoff is high. Plant growth is poor in heavily trampled ground.

To what extent do domestic livestock compete with elephants for plant resources? Elephants consumed only 3–4% of the primary production of grass in different tall grass habitats of the study area irrespective of a low or high offtake (2–42%) by domestic livestock (Chapter 6). This indicates that competition from livestock for tall grass is not important for elephants. The quality of grass during the dry season is likely to be limiting for elephants rather than the quantity of grass production in deciduous habitats. If certain habitats, such as swampy grasslands, which maintain high quality grass during the dry period, are restricted in area, there may be serious competition between livestock and elephants.

Livestock may affect elephants indirectly through their negative impact on habitat quality. A 50% offtake of grass from an area by herbivores may also cause habitat deterioration. Berwick (1976) suggests that proper use would be below 25% of production.

Livestock also transmit diseases such as anthrax, rinderpest and foot-and-mouth to wild herbivores. During 1983–86, at least 11 elephants died of anthrax in the Nilgiris.

9.5 Fire

Natural fires occur intermittently in many arid terrestrial plant communities. Still, they do not occur as frequently as human-induced fires in most regions. Fire is an established agricultural tool in many parts of the world. Shifting cultivators use fire to clear forest areas for settlement; livestock graziers set fire to unpalatable grass in order to generate a palatable flush for their animals; collectors of forest products burn the undergrowth for easy accessibility; people who frequently have to walk through a dense forest like to have better visibility to avoid elephants, some even set fire for fun or in 'vengeance against the authorities for curbing their free access to the forest'.

The study area has a long history of artificial fires (Sanderson 1878). These occur mainly between February and April during the dry season. Fires are far more frequent in the deciduous forests with tall grasses than in the short grass habitats. Most of the fires are ground fires and do not affect the canopy. The annual extent of fire was not mapped, but it was noticed that even in fire-prone zones the entire area did not burn during a given year. A particular forest patch may be burnt once every few years. An exception was Zone 2, which was burnt every year. Similarly, in the Mudumalai Sanctuary extensive fires occurred every year between 1982 and 1987.

Fire causes profound changes in the habitat. This has important implications for the survival, population size and composition of plants and animals. When evaluating the consequences of fire it is important not to subscribe to any extreme view. All fires are not necessarily destructive to the environment. Fires may help in maintaining certain desirable elements in an ecosystem, depending on management objectives. Not all ecosystems or ecological communities are adapted to fire; some communities may be undesirably altered by fire. Each situation has to be separately evaluated. Our knowledge of fire ecology in tropical forest ecosystems is inadequate in spite of fire being a burning management issue.

What are the implications of fire in elephant habitat? An immediate effect of fire is a drastic reduction in food availability. Grasses are totally burnt; the green shoots of most shrubs and young trees are also ashed, leaving only bare stems. This is generally localized in the study area because roads and artificial fire breaks usually prevent a fire from spreading for more than 5 km. Through a series of successive fires a much larger area may be burnt every dry season. Elephants simply move to adjacent areas to escape a fire. Mortality of elephants from fire is unknown.

After the onset of rains, the grass in burnt areas regenerates quickly and attracts herbivores. During the first wet season, elephants were seen actively feeding on palatable grass in burnt areas in preference to the coarse swards in unburnt areas (Chapter 5).

The effect of fire on the diversity of plant species in relation to their suitability as food is an important consideration. A regular regime of burning tends to select for plants resistant to fire. Thus, in a vegetation type which has not earlier experienced regular fires, one could expect a shift in species composition and abundance from the original climatic or edaphic climax type towards a fire-induced climax. Plant diversity may be reduced and a few fire-resistant species may dominate the vegetation. Fires may also increase the diversity of certain communities. Edroma (1981) found that fire along with moderate grazing helps maintain a high species diversity in *Imperata cylindrica* grassland. In unburnt areas the dense growth of *I. cylindrica* suppresses the growth of other species in the grassland community. Fire and grazing reduce the competitive vigour of *I. cylindrica* and allow other plants to co-exist. This does not necessarily prove that fire increases the grass species diversity, merely that it maintains the diversity in a grassland already adapted to frequent fire. A comparison has to be made with the species diversity prior to the establishment of a fire-climax vegetation in order to deduce the influence of fire on species diversity.

The dominance of the grasses *Themeda cymbaria* and *T. triandra* in the deciduous forest undergrowth in the study area is certainly the result of frequent fire. *T. triandra* is characteristic of annually burnt grasslands in Africa (Field 1976). Fire even promotes the productivity of *Hyparrhenia–Themeda* grassland in Uganda (Edroma 1984). Once these relatively coarse, perennial grasses dominate the ground layer, an annual regime of fire is needed to restore their vigour and palatability. A vicious cycle is thus set in motion, the relatively unpalatable grasses becoming dominant during the initial period of fires and, later, fires being necessary to renew their palatability seasonally.

When the tree community is considered, a characteristic feature of dry deciduous forests frequented by fire is the dominance of *Anogeissus latifolia*, which is totally useless as food for even a generalist feeder like the elephant. *A. latifolia* makes up over 50% of the tree cover in the dry deciduous zones. Pure stands of *A. latifolia* are not uncommon (Fig. 9.3).

Not all fire-adapted trees may be useless as food for elephants. Plants such as *Kydia calycina* regenerate through root coppices if fire destroys the aboveground stem. But with a regular fire regime the saplings may not be recruited into the mature tree population. This has important management implications, because elephants feed mainly on young *Kydia* trees and not on saplings (Chapter 6).

Fire acting synergistically with elephants has been responsible for converting woodland into grasslands in many regions of Africa (Buechner & Dawkins 1961; Laws 1970; Norton-Griffiths 1979). Regular dry-season fires in the Mudumalai Sanctuary for at least six years (1982–87) have resulted in sparse understorey vegetation. The size-class frequency distribution of woody plants indicates a low rate of recruitment of saplings into the mature tree class. Although rates of recruitment, growth and mortality, based on a long-term study, are needed to predict the trends in vegetation composition, the size structure is suggestive of a declining tree cover. This is not a desirable trend in the Asian elephant's habitat. *Elephas maximus* is primarily a forest species, unlike the bush *Loxodonta a. africana,* and may be unable to cope with increased heat stress caused by loss of tree cover. Any drastic increase in the proportion of grass in the diet is also not desirable (Chapter 5).

How can the elephant's forest habitat be managed for fire? If one is to be realistic, it is clear that total protection of a dry deciduous forest from fire on a long-term basis is impossible owing to the proximity of a large human

Fig. 9.3. An almost pure stand of *Anogeissus latifolia* in fire-burnt area (Zone 7). This plant is not consumed by the large mammals.

population. If an area is protected for many years, the cumulative biomass of perennial grasses will make it prone to a high-temperature conflagration that would be very destructive to the tree cover. One solution may be to resort to an early cool burn every year. Alternatively, a given block of forest could be allowed to burn once in 2–4 years, which would allow saplings to be recruited into the fire-resistant size class. Moist forests may be strictly protected from fire to prevent their degradation.

9.6 Forest plantations

The natural vegetation has been altered through artificial regeneration of native and exotic plants. The most extensive plantations in elephant habitats of southern India are those of native teak (*Tectona grandis*) and exotic eucalypts (*Eucalyptus* spp.). Other species planted to a lesser extent include silver oak (*Grevillea robusta*), bamboo (*Bambusa arundinacea*), tamarind (*Tamarindus indica*), *Acacia planifrons*, *A. ferruginea*, *Ailanthus excelsa*, neem (*Azadirachta indica*), sandal (*Santalum album*), *Albizia lebbeck*, *Pongamia glabra*, *Terminalia chebula* and *Pterocarpus marsupium*. At higher altitudes of the Nilgiri plateau are also seen extensive plantations of pines (*Pinus* spp.), Australian wattles (*Acacia decurrens*, *A. mollissima*, *A. melanoxylon*, etc.) and eucalypts.

The area under plantations in some elephant regions of southern India is given in Table 9.1. The proportion of forest area planted may be up to 45%

Table 9.1. *Area under plantation forests in some Forest Divisions of southern India*
All areas are in square kilometres.

Forest Division	Total area of Division	Area under plantations			Percentage of Division area under plantations
		Teak	Softwood[a]	Others	
Satyamangalam	1360	—	32.4	25.6	4.3
Mudumalai	321	3.8	1.1	2.3	2.2
Coimbatore	701	15.2	3.1	27.1	6.5
Chamarajanagar	335[b]	2.0	18.4	—	6.1
Hunsur	555	65.1	18.2	12.0	17.1
Nilgiris South	241[c]	—	16.9	92.5[d]	45.4

[a] The softwood plantations are mainly eucalypts and silver oak.
[b] Excludes the Gundulpet Range.
[c] Area of division within Nilgiri Biosphere Reserve.
[d] Mainly wattle plantations.

the area of a division. Plantations occupied about 7% of the main study area in 1981. The most commonly planted species is *Eucalyptus*, followed by *Tectona grandis* and *Grevillea robusta*.

Planting of bamboo and certain species of *Acacia* has increased the food resources, though not to any appreciable extent. *Acacia planifrons* planted in the plains (Zones 18 and 19) is not yet consumed by elephants, although other species of *Acacia* with a similar growth form are eaten. The bark of *Tectona grandis* is eaten to some extent. The bark of *Eucalyptus*, which had not been consumed during the early years of its introduction, has not become part of the elephant's diet. This testifies to the elephant's ability to partially adapt to changing habitat conditions.

Ultimately, most monoculture plantations may be inferior to natural vegetation in terms of availability of food plants. The most important attribute is the nature of the undergrowth, upon which elephants feed, in the plantation. To achieve the best growth rates of plantation trees the undergrowth is usually cleared. This drastically reduces the availability of young trees and shrubs for animals, although some grass cover may be retained.

Clear-felling of natural vegetation on hill slopes to raise plantations results in loss of top soil. The undergrowth in many plantations is also invaded by useless weeds such as *Lantana camara* and *Chromolaena odorata*, which not only suppress the growth of the planted trees but also reduce food availability.

In the study area, the silver oak plantations had a very poor undergrowth. Because this plant is not consumed by elephants, such plantations were clearly useless habitat for elephants. *Eucalyptus* plantations were raised in the Satyamangalam Division in areas where planted bamboo had failed owing to pressure from livestock, elephants and fire. The undergrowth in these plantations had bamboo and other food plants. *Eucalyptus* planted in gaps in scrub vegetation also, perhaps, does not lower the habitat quality. The process of raising a *Eucalyptus* plantation in natural vegetation by clearing the undergrowth of shrubs and leaving only the mature trees is not desirable, since elephants prefer to feed on younger plants.

The frequency of woody food plants in natural forest compared with teak plantations was enumerated in two regions (Table 9.2). In both regions the natural vegetation had a higher availability of plants such as bamboo, *Kydia calycina*, *Grewia tiliaefolia* and *Helicteres isora* in the preferred girth class of 10–60 cm. High herbivore densities may be maintained in teak plantations if there is sufficient growth of grasses as in the Nagarhole National Park (Fig. 9.4).

Table 9.2. *Food plant availability in teak plantation versus natural forest*

	Number of plants per hectare			
	Mudumalai Sanctuary		Chamarajanagar Division	
	Teak plantation	Natural forest	Teak plantation	Natural forest
Teak plants				
10–60 cm girth	78	5	200	0
>60 cm girth	284	58	98	0
Other food plants[a]				
10–60 cm girth	8	158	174	594
>60 cm girth	8	28	10	30
Bamboos				
Number of culms	114	158	b	b

[a] Other food plants are mainly members of the order Malvales.
[b] Only negligible quantities of regenerating bamboo.

Fig. 9.4. A herd of gaur in a teak (*Tectona grandis*) plantation in the Nagarhole National Park, southern India. A high biomass of herbivores is maintained in the plantations in this park.

9.7 Food availability in primary versus secondary rain forest

The most detailed survey comparing availability of food plants for elephants in primary and secondary rain forest is that by Olivier (1978b) conducted in peninsular Malaysia. The number of 'individual' food plants (trees, palms, herbs, bamboos and grasses) were enumerated in a large number of plots of 400 m² each and the results converted into 'trunkfuls' after suitably weighing the numbers according to size classes (Table 9.3). The primary forests sampled had no history of any logging; the secondary forests included those which had regrown after complete clearance of primary forest and those which were primary forest degraded through logging.

The secondary forests had an average of 214 trunkfuls of food plants per plot, compared with 157 trunkfuls in primary forest. The increased food availability in the secondary forest was mainly from trees of small and medium size classes. The opposite was true of the palms which form the major food category. Food available from small palms was considerably greater in the primary forest, whereas medium and large palms showed higher values in the secondary forest.

9.8 Carrying capacity of primary versus secondary vegetation

It has often been observed that secondary habitats, including former sites of human settlement, have a higher density of herbivorous mammals. In the Serengeti National Park, Tanzania, the impact of man through burning of the grassland and grazing of domestic livestock is

Table 9.3. *Food plant availability in primary and secondary rain forest*

Plant category	Mean number of individual food plants per plot		Mean weighted trunkfuls of food plants per plot	
	Primary forest	Secondary forest	Primary forest	Secondary forest
Trees	8.6	38.5	19.0	71.4
Palms	79.6	52.5	103.3	124.5
Herbs	14.0	11.3	15.1	14.4
Bamboos	1.3	0.3	19.1	1.8
Grasses	0.1	2.3	0.1	2.3
Total	103.6	104.9	156.6	214.4

Each plot is 400 m² in area.
Modified from Olivier (1978b).

long-established. The highest animal densities are found in such areas of past and present pastoral activity (Bell 1971). Some high-density elephant habitats in southern India were formerly sites of settlement.

The question is whether or not a high elephant density is directly due to human activity creating a more favourable habitat? Sites of human settlement are usually prime habitats with abundant water supply, fertile soil and high potential productivity. These would certainly have been attractive to elephants even before human occupation. Elephants might be reasserting their former levels of abundance after the sites are abandoned by people.

Nevertheless, the conversion of certain climax vegetation types, such as the tropical evergreen forest, to secondary forms by human activity results in an increase in elephant density. Olivier (1978b) proposed a model depicting an increase in elephant density along two gradients: one gradient across forest formations, from evergreen forest (low carrying capacity K and elephant density D) to deciduous forest (high K and D), and another gradient within forest formations, from closed canopy climax forest (low K and D) to open canopy seral forest (high K and D).

These trends hold good only up to a point. A modified version of the model is shown in Fig. 9.5. Density figures are approximately the levels seen in southern Indian habitats. Elephant density increases from an evergreen forest through the semi-evergreen type only until the deciduous forest. Beyond this there is a decline as the vegetation changes progressively into

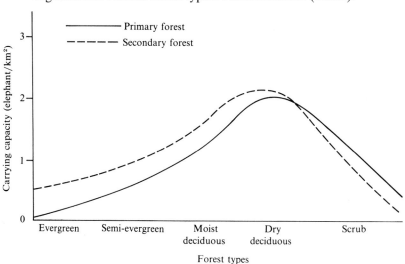

Fig. 9.5. A model of carrying capacity of primary versus secondary vegetation in various forest types. From Sukumar (1986a).

poor scrub. The model does not take into consideration any major differences in water availability. A scrub habitat along a perennial river may support a higher density than a dry deciduous forest with scarcity of water.

A change from climax stage to early seral stage also does not elicit the same response in different vegetation types. The highest relative increase in carrying capacity occurs in the evergreen forest, but beyond the moist deciduous forest it is unlikely that this trend will continue. A dry deciduous forest or xerophytic vegetation is characteristically short-statured, ensuring a high proportional browse availability for elephants. Further alteration of dry scrub vegetation through human exploitation would reduce the carrying capacity.

9.9   Creation of water reservoirs

Artificial water bodies, from small ponds to large reservoirs, are very common in elephant habitat. Ponds are dug inside the forest for domestic livestock taken for grazing, or may be the outcome of management policies for a reserve. Dams impound reservoirs of various sizes, shapes and volumes in order to generate hydro-electric power or provide water for irrigation. In southern India there are over 40 dams creating water bodies with submersion areas varying from 5 to 150 km^2 within elephant habitat.

Ponds and dams may serve to localize elephants for longer periods than their 'natural' movement pattern would allow. Such concentrations of elephants in Africa are known to destroy woodlands in the surrounding areas (see, for example, Corfield 1973). In the Bandipur National Park of Karnataka state, the presence of numerous artificial ponds within a small area invites intensive utilization by elephants. Fears of a decline in tree cover have been expressed but this has not been quantitatively confirmed.

Dams result in a loss of prime river valley habitats for elephants. Because elephants utilize river valleys in greater proportion than their availability, the loss in habitat area and food supply will be more serious than simple statistics of submerged area reveal. Reservoirs boost the availability of water but this will not necessarily benefit elephants to an appreciable extent. In the first place, large reservoirs have been created on perennial rivers, which in any case would have provided sufficient water for their needs. Small dams which impound seasonal streams may, however, significantly increase the availability of water which may be crucial for elephants during periods of drought. In that sense, artificial water bodies would lead to the maintenance of elephant populations at higher densities than the natural habitat would support.

Dams and their associated developments have also disrupted the traditional

160 *Habitat manipulation by people*

movement of elephants or caused an indirect reduction in habitat by increasing the extent of cultivation. The impact of two major dam projects, the Parambikulam–Aliyar Project in the Anamalai hills of southern India and the Accelerated Mahaweli Ganga Project in Sri Lanka, on elephant populations is described below. A number of lessons can be learnt from these examples.

9.9.1 *The Parambikulam–Aliyar Project in southern India*

In the Anamalai hills of Kerala and Tamilnadu states, a series of dams and their associated canals and pipes have affected the movement of elephants (Fig. 9.6). The reservoirs, with submerged areas varying from 0.1 to 21 km^2, range in altitude from 250 to 1200 m above mean sea level.

Fig. 9.6. Map showing the reservoirs and canals of the Parambikulam–Aliyar project in the Anamalai hills, southern India.

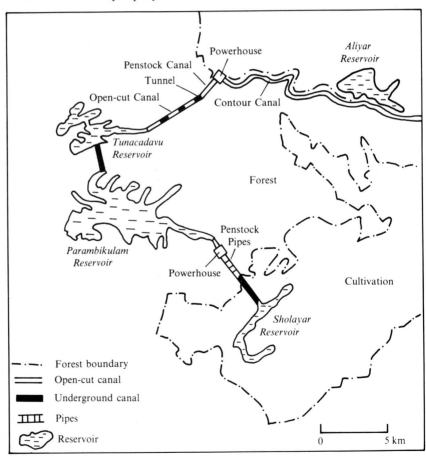

Reservoirs at higher altitudes on the hills are used to generate electricity, while those at lower altitudes along the periphery of the forest divert water into irrigation canals. There has been no agricultural development associated with these reservoirs in the hill forests, which constitute the Anamalai and Parambikulam Sanctuaries.

Obstacles to elephant movement are seen in many places. The waters of the Sholayar reservoir are taken to a power house through penstock pipes laid along the steep hill slope. These pipes prevent elephants from moving across the slopes. The canals connecting the reservoirs are partly underground and partly above ground. The canal leading from the Tunacadavu reservoir to the Sirkarapathy power house is open-cut along four stretches (Fig. 9.7). Many animals, including elephant, gaur, spotted deer, sambar and even a tiger, have been washed away by the swift current when they have attempted to cross the canal. These animals usually end up at a tunnel entry. Ramps constructed at tunnel openings prevent animals from being sucked inside, but

Fig. 9.7. An open-cut canal between the Tunacadavu reservoir and the Sircarapathy power house in the Parambikulam–Aliyar project. Elephants and other animals have often been washed away and drowned in the canal.

the animals are stranded unless rescued after considerable effort. At least 10 elephants have died in these canals, and another 15 have been rescued since 1970. Rescue involves stopping the flow of water and using domestic elephants to lift the trapped elephants. The last stretch of the canal going downhill to the Sirkarapathy power house is again an impediment to movement. The feeder canal from the power house to the Aliyar reservoir along the forest boundary has served the useful purpose of preventing elephants from entering cultivation.

9.9.2 The Mahaweli Ganga Development Project in Sri Lanka

In 1979, Sri Lanka launched the major Accelerated Mahaweli Development Programme. This involves the construction of dams across the country's largest river for generating electricity and providing irrigation for additional agricultural lands. By 1986, the major dams that had been completed were the Victoria, Madura Oya, Kotmale and Randenigala, which together submerged about 120 km^2 of forest and existing agricultural land. Two smaller reservoirs, Ulhitiya and Ratkinda, were also created, while work was yet to commence on the Rantembe dam. A network of underground tunnels and open canals link the reservoirs and divert water for irrigation. Up to 3600 km^2 of agricultural land were to be developed and a million people settled in the project basin.

Estimates of the area of new forest land and wildlife habitat to be cleared for agricultural development vary from 1250 to 2600 km^2. About 800 elephants inhabit the Mahaweli basin at a pre-project density of 0.17 elephant/km^2. The loss of habitat would eventually compress them to a density of 0.38 elephant/km^2 or more than twice the initial density. The Mahaweli Environment Programme has set up a system of protected areas for wildlife conservation within the Mahaweli basin (Fig. 9.8). The reserves already created or proposed include the Somawathiya National Park (210 km^2), Wasgamuwa NP (338 km^2), Flood Plain NP (174 km^2), Maduru Oya NP (515 km^2), Trikonamadu Nature Reserve (250 km^2), Minneriya–Giritale NR (420 km^2) and the Nilgala Jungle Corridor (280 km^2). These are further connected with the extensive Hurulu Forest Reserve in the north and with the Gal Oya NP in the south. There is a major discontinuity in habitat between the Wasgamuwa NP and the Maduru Oya NP. Ishwaran & Banda (1982) proposed the Kuda Oya corridor linking these two parks, but this seems to have been rejected on economic grounds. It must be pointed out that only a fraction of the area under the reserves are under forest cover (30% for Maduru Oya NP), the rest being grasslands or abandoned *chena* (shifting agriculture) land.

Fig. 9.8. Map of the Mahaweli Development Project area in Sri Lanka showing the reservoirs, agricultural areas and protected areas. Areas A–G are agricultural developments; NP, National Park; NR, Nature Reserve.

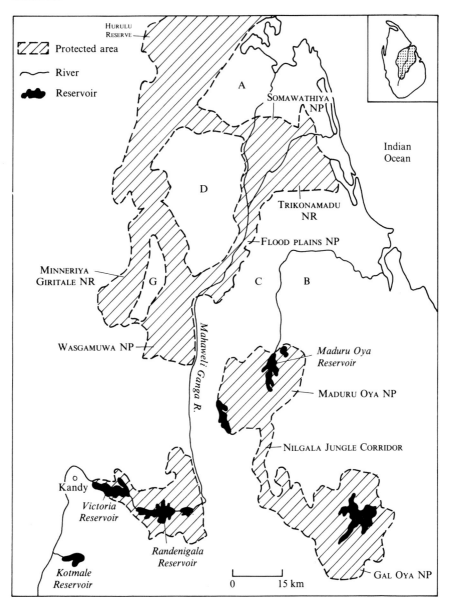

The impact of the Mahaweli Project on the elephant population will be known in the coming years. The reservoirs will submerge high-quality river valley habitat. Seasonally inundated marshy areas, known as *villus,* in the Mahaweli flood plain are expected to decline to half their original extent. The grassy *villus* are important feeding grounds for elephants. Fragmentation of habitat will increase the conflict between compressed elephant herds and agriculture. The long-term viability of elephant populations may decrease.

10

Elephant slaughter by people

Apart from the regular capture for domestication, the elephant populations of Asia have also been severely reduced as a result of hunting male elephants for their tusks and killing of both sexes in defence of agricultural crops. After the enaction of wildlife protection laws, the second factor has probably reduced, but the illegal hunting for ivory continues unabated at least in southern India.

Unlike the Asian elephant, both male and female African elephants possess tusks. Thus, both sexes are affected by hunting for ivory, although the males suffer a proportionately higher mortality because of their larger tusks. The impact of hunting on African elephant populations has been much debated. Douglas-Hamilton (1980) reported that between 50 000 and 150 000 elephants fell prey every year to ivory hunters, resulting in declining populations. Parker & Martin (1982) estimated that less than 50 000 elephants contributed to the annual African ivory trade since 1977. They concluded that this level of harvest was below the annual reproductive potential of the total African elephant population. More recently, Douglas-Hamilton (1987) presented further evidence for a decline in most African elephant populations due to hunting.

This chapter examines the role of hunting in elephant mortality and its contribution to the ivory trade in India. The impact of hunting on elephant population dynamics will be discussed in Chapter 11.

10.1 Elephant populations susceptible to poaching

The proportion of male elephants possessing tusks varies enormously among different Asian elephant populations (Table 10.1). Tuskless bulls are locally known as *makhnas, aliyas,* etc. (Fig. 10.1). Fewer than 10% of elephant bulls in Sri Lanka are tuskers. This has conferred upon them a virtual immunity against poaching. In northeastern India there are approximately equal numbers of tusked and tuskless bulls. Poaching for ivory occurs in these states but quantitative estimates are not available. Poaching for meat occurs in the states of Mizoram and Nagaland. In southern India, where over 90% of the bulls are tuskers, poaching is a serious conservation issue.

Elephant populations in continental Southeast Asia also seem to have a high proportion of tuskers but sufficient data on the extent of poaching are not available. Hunting occurs in Burma and Thailand, but not in peninsular Malaysia. The returns from hunting may be inadequate considering the effort needed to locate tuskers in regions of very low elephant density.

10.2   Causes of death in elephants

The proportions of elephants dying of natural causes and of those killed directly by people in southern India are given in Table 10.2. One difficulty in getting a correct picture is that instances of poaching are under-represented in the official records. The extent of this bias varies from one state to another, among Forest Divisions within a state and from one time period to another. The figures for 1975–87 have been corrected, with

Fig. 10.1. A tuskless male elephant in Yala National Park, Sri Lanka.

Table 10.1. *Incidence of tusklessness in male Asian elephants*

Region	Year of sampling	Sample size	No. of tuskless bulls	% tuskless in male population
1. Madras Presidency & Tamilnadu (S. India)	1926–80	217 captured	4	1.8
2. Travancore–Cochin (S. India)	1939–51	65 captured	0	0
3. Karnataka (S. India)	1968–74	55 captured	3	5.5
4. Satyamangalam–Chamarajanagar (S. India)	1981–83	24 adult wild	2	8.3
5. Assam (N.E. India)	1937–50	1313 captured	639	48.7
6. Assam (N.E. India)	1981	67 captured	31	46.3
7. Sri Lanka	up to 1955	c. 364 captured	324	89.0
8. Lahugala (Sri Lanka)	1967–69	25 adult wild	23	92.0
9. Peninsular Malaysia	up to 1954	43 shot	1	2.3

Sources of information: 1 and 3, from Forest Department records; 2, 5, 7 and 9, Deraniyagala (1955); 4, registered elephants during the present study; 6, Forest Department census; 8, McKay (1973).

Table 10.2. *Causes of death in elephants in southern India*

| | Female elephants | | Male elephants | | | |
	Natural death	Crop defence	Natural death	Crop defence	Attempted poaching	Poaching
Tamilnadu						
1975–87	77 (80.2%)	19 (19.8%)	103 (37.2%)	29 (10.5%)	21 (7.6%)	124 (44.8%)
1981–82	14 (66.7%)	7 (33.3%)	23 (26.1%)	4 (4.5%)	1 (1.1%)	60 (68.2%)
Karnataka						
1975–87	85 (85.0%)	15 (15.0%)	61 (32.4%)	8 (4.3%)	21 (11.2%)	98 (54.1%)
1981–82	20 (95.2%)	1 (4.8%)	17 (27.9%)	0 (0.0%)	8 (13.1%)	36 (59.0%)
Total						
1975–87	162 (82.7%)	34 (17.3%)	164 (35.3%)	37 (8.0%)	42 (9.0%)	222 (47.7%)
1981–82	34 (81.0%)	8 (19.0%)	40 (26.8%)	4 (2.7%)	9 (6.0%)	96 (64.4%)

The records for Karnataka pertain only to 5 Forest Divisions. The records for both states are complete only for 1981–82.

168   *Elephant slaughter by people*

additional male deaths by poaching inserted only for 1981–82, based on personal knowledge of such unreported cases. Thus, the overall results still tend to underestimate the role of poaching. The corrected results for 1981–82 are also presented separately.

The pattern of mortality did not differ much between Karnataka and Tamilnadu states. While poaching (tusks stolen) accounted for 48–64% of male elephant deaths, another 6–9% were found dead in the forest with gunshot wounds but with their tusks intact. Most of them seemed to have died in attempted poaching, although it is possible that some had been shot while raiding crops. About 3–8% were certainly killed in defence of crops. Overall, people were responsible for 65–73% of male elephant deaths, the rest dying of natural causes.

People also caused 17–19% of female elephant deaths by shooting or electrocuting them in defence of crops.

10.3   Number, age frequency and mean tusk weight of poached elephants

(a) Numbers killed

It was estimated that 30–50 male elephants were killed by poachers each year in each of the three states of Kerala, Karnataka and Tamilnadu between 1977 and 1986. Thus, a total of about 100–150 bulls were killed each year in southern India. The tusks were recovered in only 10–20% of the cases. At least 100 pairs of tusks were lost to poachers annually.

For the above period the most complete official records are available for Karnataka and the least reliable for Kerala. It is common knowledge that poaching was rampant during the 1970s in the Periyar Tiger Reserve of Kerala. Owing to a depletion of tuskers in Kerala, poachers shifted their attention to Tamilnadu and Karnataka during the 1980s. In these two states the forest regions seriously affected include Mudumalai, Nilgiri North, Satyamangalam, Erode, Hosur, Hunsur, Bandipur, Chamarajanagar and Kollegal.

(b) Age frequency of poached bulls

The age class frequency of a sample of 95 male elephants poached in the Nilgiri – Eastern Ghats region during 1979–83 is given in Chapter 11. The living age structure of males for this larger region is not available, but a comparison has been made with the age structure of males in the main study area. Bulls are vulnerable to poaching from age 5 years onwards. Those in the 5–10 year age class were poached to a lesser extent than their availability in the population. For males above 10 years there seemed no clear selection for any age class. Poachers seemed to shoot them roughly in the same proportion as they encountered these elephants. Owing to a decline in the number of large bulls it is likely that these can be located only after considerable effort; hence, poachers tend to shoot whatever they come across.

(c) Mean weight of poached tusk

The mean weight of a poached tusk was calculated in the following manner. Records of tusks retrieved from dead elephants in Karnataka between 1965 and 1983 were considered. Since poachers do not generally shoot male elephants below 5 years, only tusks above 15 cm circumference at the lip line or weighing 1 kg (corresponding to age 6 years) were used in the calculation. A sample of 247 tusks gave a mean tusk weight of 8.6 kg.

For the sample of 95 poached elephants whose ages were estimated, the tusk weights were calculated from the growth curve of tusk weight with age (Appendix II). This gave a mean tusk weight of 10.6 kg. Because the growth in tusk weight curve refers to unbroken and unworn tusks, the actual weight retrieved by the poachers would be slightly less. Based on these calculations the mean weight of a poached tusk can be taken as approximately 9.5 kg.

10.4 The people responsible

The people who shoot or electrocute elephants raiding their crops are usually the wealthier farmers. The ivory poaching organization has at least three links: the person who shoots the elephant and extracts the tusks, the middleman who handles the tusks in transit and the ivory dealer who purchases the tusks for further processing.

The people who actually kill the elephant and remove the tusks hail from villages inside or adjacent to the forest. The hunters may be organized into a large group (up to 30 people have been seen in the study area and 200 in the Periyar Reserve) or a number of small groups all catering to the same middleman. They may spend most of their time inside the forest, shift their camps regularly and return to their villages only occasionally. The hunters may combine ivory poaching with sandalwood and timber poaching.

The smaller, ill-organized groups shoot with only muzzle-loading guns; the better organized poachers may use modern automatic rifles. Poachers often keep watch at a water hole for elephants to appear and then track them into the forest before shooting them. Some poachers have even used trained dogs to track elephants and separate sub-adult males from the herd. Once an elephant is killed the extraction of the tusks may take less than three hours (Fig. 10.2). Tusks are carried out of the forest on foot. For their work the hunters receive only a fraction of the market value of ivory.

The middleman is the crucial link in the illegal ivory trade. He is usually a middle-class man residing in a small town close to the forest. He organizes the hunters and collects the booty. He maintains contact with ivory dealers in the larger cities. The ivory dealer who finally purchases the tusks from the middleman may be a licensed dealer who is willing to use loopholes in the law to purchase local ivory, or a smuggler who illegally exports the ivory.

10.5 Poaching and the ivory trade in India

There are about 7200 ivory craftsmen in India, mainly in Kerala (3000), Delhi (2000), Jaipur (800) and Mysore (600). Thus, at least 50% of the carvers in the country are in southern India. Carvers in the south work mainly by hand as opposed to those in the north who use electric lathe machines. Output of finished ivory products may be less in southern India, but these are of much higher quality and value (Martin 1980).

Ivory from Indian elephants has never been sufficient to sustain the demands of the carvers. As early as the sixth century BC, African ivory from Ethiopia was being sent to India (Warmington 1974). African ivory has been imported in large quantities from the nineteenth century. During 1875–81 about 250 tonnes of ivory were imported every year. During 1940–45 the

Fig. 10.2. Remains of a male elephant whose tusks have been poached.

annual imports were roughly the same, but from 1948 onwards the quantity declined substantially. During 1960–71 an average of only 37 tonnes were imported yearly. The main reason for this decrease was the high import duty (over 100%) on ivory, which the merchants in India cannot afford. There were not many restrictions on ivory imports until 1977, when the elephant was included in Appendix II of the Convention on International Trade in Endangered Species of Wild Fauna and Flora (CITES). By 1979 the imports had come down to 5 tonnes (Prasad 1981). Current annual imports are less than 15 tonnes.

The supply of legal Indian ivory to the carvers has always been lower than that of imported ivory. In 1978 the ivory (from elephants found naturally dead) sold by the Forest Departments of all states in India amounted to only 3 tonnes (Martin 1980). Between 1974 and 1980 the average quantity sold or auctioned by the southern states came to about 1 tonne per year. Sales by Karnataka were entirely to the state-owned Handicrafts Board. In 1982 the auctioning of ivory was stopped by Tamilnadu. All trade in Indian ivory was finally banned in November 1986.

Over 85% of the finished African ivory is sold to foreigners. This includes legal exports to various countries. In 1979 the total exports were 12.95 tonnes, which went mainly to Germany, Italy, France, U.S.A., Denmark and the Netherlands. A certain unknown quantity is also taken out as accompanied baggage by tourists. Ivory is also smuggled out to the Arab countries.

The price of raw Indian ivory within the country has gone up enormously in recent years (Table 10.3). This parallels a similar upsurge in the world market price of raw ivory. Between 1870 and 1930 the ivory price index was generally steady and parallel to the gold index, pound sterling index and the commodity price index (Fig. 10.3). It continued to be steady from 1930 to 1970 and lagged far behind the other indices, which showed an exponential increase. Beginning from 1970 or so, the ivory index shot up with a vengeance and by 1978 had caught up with the other indices. People had suddenly realized that raw ivory could be stored as a profitable investment in order to counter inflation. In absolute terms, however, the annual legal exports of Africa's raw ivory declined from 991 tonnes in 1976 to 680 tonnes in 1980. About 83% of this quantity was exported to Hong Kong and Japan (Parker & Martin 1982). Large quantities are also purchased by the Arab countries.

Compared with the supply of African ivory to the world trade, the ivory from Indian elephants is a pittance. Why is there a demand for local ivory? Although the ivory craft is on the decline, there are more carvers in India than in any other country. A regular supply of ivory is necessary to sustain their

Table 10.3. *Raw ivory prices in the Indian and world markets*
Prices are in dollars per kilogram.

Year	In India	In the world market
1966	14	6
1970	—	7
1976	52	34
1978	—	74
1980	130	70
1982	150	53
1984	136	46
1986	120	100
1989	200	190

The apparent decrease in the price of Indian ivory after 1982 is due to the weakening of the Indian rupee in relation to the US dollar from INR9.50 per dollar in 1982 to INR12.50 per dollar in 1986 (1 dollar = INR15.25 in 1989).

profession. Local ivory from poached elephants becomes attractive as it is cheaper than the prevailing imported African ivory. Ivory dealers can usually pass off finished products made from Indian ivory as coming from African ivory which they have legally imported.

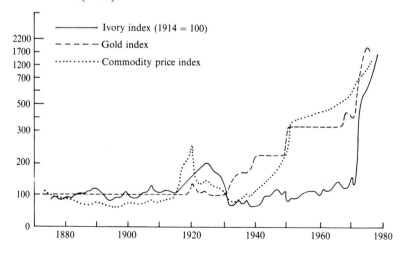

Fig. 10.3. Trends in the ivory index in relation to other indices. Source: Jackson (1982).

What is the value of the illegal trade in tusks from poached elephants? It has been already mentioned that from 1977 until 1986 at least 100 tuskers were shot every year by poachers in southern India. This would have given an annual yield of 190 tusks (assuming that some elephants were one-tusked). With a mean tusk weight of 9.5 kg this would amount to 1800 kg. At the 1982 price of $150 per kg, the total annual value of the poaching trade is $270 000. (This should be taken as a conservative estimate, as the true figures for poaching in Kerala were not available.)

Sri Lanka is known to export carved ivory although it does not import African ivory. The source of this ivory is not known. Burma exported 5.13 tonnes of local ivory to Japan during 1978–86. The tusks may have come from elephants which died of natural causes. In addition, the poaching trade in continental Southeast Asia may be handling larger quantities of ivory.

In 1988 the Indian government abolished the customs duty on ivory imports. By then, however, a large proportion of ivory dealers in India had stopped importing African ivory. A preliminary survey in 1989 indicated that there were only about 2000 ivory carvers in India (compared with 7200 carvers in 1978), still higher than in any other country (E. B. Martin, personal communication).

11

Population dynamics

One of the aims of this study was to determine the impact of human–elephant interaction on the dynamics of the elephant population. For conserving a species it is important to know not only the current trends in population numbers – whether the population is increasing, stable or declining – but also other demographic parameters, such as age structure, sex ratio, fertility, mortality and physiological condition, which would ultimately influence the size and viability of a population.

The dynamics of a mammalian population are invariably linked to habitat conditions such as availability of food and the presence of competitors and predators. Thus, interactive models incorporating second-order effects (age structure, competition, plant succession, etc.) are improvements over simple single-species models (Caughley 1981). Such models must also ideally contain a measure of resilience or the capacity of the population to bounce back to its original state after a setback (Hanks 1981). The more complex models require information that is rarely available for wild populations.

Various population models have been constructed for the African elephant (Hanks & McIntosh 1973; Fowler & Smith 1973; Wu & Botkin 1980; Croze et al. 1981). These models have deduced how variations in juvenile and adult mortality, age at first calving and inter-calving interval influence r, the instantaneous growth rate of the population. Fowler & Smith (1973) further incorporated density-dependent changes in these parameters into their Leslie matrix model. Wu & Botkin (1980) have come up with a stochastic model considering variations in environmental conditions and differences in individual life-histories.

In this study it was not possible to obtain the detailed information on elephant reproductive biology that is available for African elephant populations from examination of culled elephants. Estimates of fertility and mortality were obtained during 1981–83 by simple field methods and incorporated into an age-structured Leslie matrix model. Trends in various demographic parameters were simulated using a computer. Details of the Leslie model are available in many texts on population biology (e.g. Pielou 1977). The aspects of population dynamics of primary interest for the elephants of southern India were the following.

(a) Starting with the current age structure and fertility schedule, how does variation in the mortality schedules for male and female elephants influence the population growth rate? What is the population growth rate given the currently prevailing mortality rates? What is the maximum growth rate under the best possible conditions?

(b) Since male mortality is far higher than female mortality due to both natural causes and hunting by man, how does the sex ratio change with different mortality schedules?

(c) How do the demographic parameters in the study population compare with those in other African and Asian elephant populations?

Evidence that the model was reasonably robust came from observations on the same population during 1987.

11.1 Age and sex structure

The age structure of elephant family groups sampled during 1981, 1982 and 1983, and the average age structure of adult males, are given in Table 11.1. An average three-year age structure of the elephant population incorporating both family groups and adult bulls is shown in Fig. 11.1.

During any one year the pattern of peaks and troughs in the lower age classes indicates differences in annual recruitment (Laws et al. 1975). It was not possible to sex elephants below 2 years; an equal sex ratio at birth has therefore been assumed. Evidence for this comes from records of 187 elephants (97 males and 90 females) born in captivity in southern India. In the 3–5 age class of the wild population there were more males than females, but in no year did this depart significantly from a 1:1 ratio (χ^2 test, $p > 0.10$). From the 5–7 age class onwards there was a sharp reduction in the number of males as a result of higher mortality from poaching. The male segment seems to have an unstable age distribution due to the irregularities in mortality caused by poaching. The female segment, however, shows a more stable age distribution. Peaks in the 5–7 and 15–20 age classes may have arisen partly from slight inaccuracies in the ageing technique (Hanks 1972a) and also from

176 Population dynamics

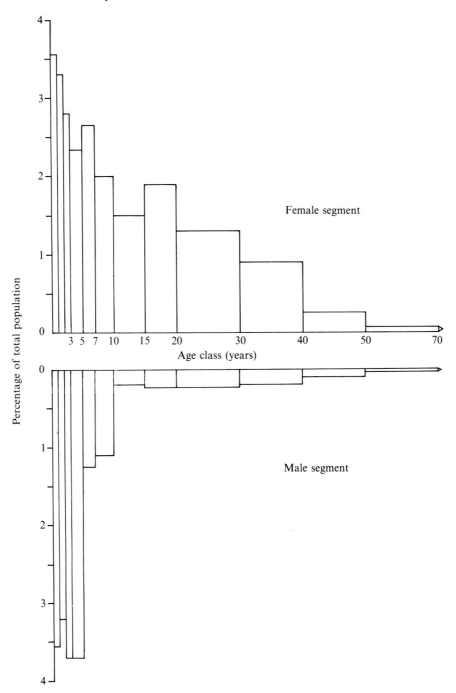

Fig. 11.1. Average age structure, for 1981–83, of the elephant population in the study area.

Fertility

Table 11.1. *Annual age structure of the elephant population*

Age class			Number of elephants in each age class						Combined 1981–83		Percentage of elephants in the age class	
			1981		1982		1983					
Category		Years	F	M	F	M	F	M	F	M	F	M
Calf		0–1	8		13		2		23		3.6	3.6
Juvenile	I	1–2	6		8		7		21		3.2	3.2
	II	2–3	4	4	2	3	3	5	9	12	2.8	3.7
	III	3–5	6	9	6	12	3	3	15	24	4.6	7.4
Sub-adult	I	5–7	5	2	9	4	3	2	17	8	5.2	2.5
	II	7–10	8	5	7	5	4	1	19	11	5.9	3.4
	III	10–15	10	0	10	3	4	0	24	3	7.4	0.9
Adult	I	15–20	11	—	11	—	9	—	31	4	9.6	1.2
	II	20–30	15	—	17	—	10	—	42	8	13.0	2.5
	III	30–40	10	—	12	—	7	—	29	7	9.0	2.2
	IV	40–50	4	—	2	—	2	—	8	4	2.5	1.2
	V	50–70	1	—	2	—	1	—	4	1	1.2	0.3
Total			108		126		66		324		67.9	32.1

Calves below 2 years could not be sexed. Combined 1981–83 age structure for adult males is presented. The proportion of adult males in the population was estimated to be 7% from a larger sample (details in text).

annual differences in recruitment and chance deviations from an equal sex ratio at birth.

In a larger sample of 1331 elephants that were sexed and classified during 1981–83, the 89 adult male sightings constituted 6.7% of the population. The 24 registered bulls constituted a slightly higher proportion (7.5%) of the sample of 324 elephants used for the computation of age structure. Thus, the proportion of adult males was correspondingly reduced in the initial age structure used for the simulations.

The incidence of tusklessness among male elephants is quite low throughout southern India. Only 2 (8.3%) out of 24 registered adult bulls in the study area were tuskless bulls or *makhnas*. Since tuskers have a higher mortality rate from poaching, it is possible that the percentage of *makhnas* among juveniles would be still lower. It was difficult to identify young *makhnas* among family groups.

11.2 Fertility
11.2.1 *Age at first and last calving*

Onset of puberty or first oestrus in mammals may be influenced by nutrition and environmental factors (Sadleir 1969). The elephant is physiologically capable of conceiving at the time of sexual maturity or first ovulation, but it may not do so for various reasons (Laws *et al.* 1975). For the purpose of demographic change, the mean age at first conception and the age at first calving are important. The mean gestation period is between 20 and 22 months in the Asian elephant.

Records of elephants kept in captivity provided useful information on the age at first calving, although this should not be necessarily taken as representative of wild populations. Among domestic elephants in Burma the earliest age of calving recorded by Burne (1942) was 16 years. Among the 88 elephants captured in the Mysore Kheddah during January 1968, there were six cows in the 10–15 age class, 10 cows in the 15–20 class and 18 cows above 20 years. Out of nine identified 'mother – calf below 1 year' units, only one mother came in the 10–15 age class (it was 12 years as estimated by height), three mothers were in the 15–20 age class and the rest were above 20 years. Five more cows gave birth within ten weeks of capture. Of these the youngest mother was about 15 years old. It was not possible to verify whether some of these 14 mothers with calves below 1 year had given birth a second time.

The records of over a thousand elephants captured in southern India or born in captivity were examined. The earliest age of calving was 13.3 years for an elephant named Meenakshi which itself was born in captivity. A few cows gave birth between age 14 and 18 years, but the majority of them calved only after 18 years. In captive Asian elephants it must be exceptional for a cow to give birth before 12 years. The mean age at first calving, however, seems to be much later, probably between 18 and 20 years.

The mean age at first calving for the wild population was estimated following the method of Douglas-Hamilton (1972). From the identified family units it was determined whether or not each cow elephant above 10 years had any offspring. Mothers of calves were identified based on their lactational status and nearest neighbour relationship. The data, grouped into three age classes, are given in Table 11.2. There was only one cow below 15 years which had a calf, while there were two cows above 20 years which did not have offspring or swollen mammary glands. Since most cows between 20

Table 11.2. *Reproductive status of female elephants*

Age class (years)	Number of females in sample	Number of females which had calved	Percentage of females which had calved
10–15	12	1	8.3
15–20	16	9	56.3
20+	48	46	95.8

and 25 years seemed to have calved prior to 20 years, the data suggest that the mean age of first calving is between 15 and 20 years. A likely source of error is that some of the calves born to cows in the younger age classes may have died. Given all these limitations it is reasonable to take the mean age of first calving for the wild population as 17–18 years.

Elephants are known to produce calves up to an advanced age. In African elephants the age of menopause has been estimated to be 55 years (Laws *et al.* 1975). Smuts (1977) even recorded a 60-year-old cow that was pregnant.

Among captive Asian elephants, the cow Meenakshi (which died at the age of 68 years) gave birth to its last calf at age 54 years, while another cow, Tara (still alive in 1989 at estimated age 75–80 years), last calved when it was at least 62 years old. The largest and oldest females in the wild population were reproductively active. Two cows estimated to be definitely over 50 years had calves aged three years and one year when first observed. The exact age of menopause is not important in elephant population dynamics since there will be only a very small proportion of old cows in the female segment.

Sexual maturity in male elephants could not be precisely estimated. Laws *et al.* (1975) have shown that the mean age of sexual maturity in both male and female elephants is the same for a particular population, although it may differ from one population to another. Bulls in the study population became completely independent of their families by the age of 15 years. Some bulls in the 15–20 age class were seen in musth, an indication of sexual maturity. Bulls may not be able to mate until older than 20 or 25 years owing to the prevailing social hierarchy. In considerations of population dynamics, the age of sexual maturity in males is generally not relevant except when there is an abnormal depletion of older bulls. In such a situation the younger adult males may contribute to successful conceptions.

11.2.2   *Birth rate and the mean calving interval*

The birth rate of a population is an important attribute of its demographic vigour. Ideally, the age-specific fertility should be known, but usually only a mean calving rate can be determined for the entire mature female population. Commonly, the percentage of calves below one year in the sampled female population has been used as a measure of fertility.

Too simplistic an interpretation of such data based on cross-sectional sampling can lead to serious errors due to many reasons. There may be large annual fluctuations in the recruitment to the population. Laws *et al.* (1975) found that the percentage of female elephants pregnant at a given time may vary widely from one population to another. Although they were not able to sample the same population during successive years, they have stressed the possibility that such annual fluctuations in the rate of conception may occur in a population. This was inferred from the age structures which showed peaks and troughs, indicating cycles in recruitment with a period of 6–8 years in response to environmental fluctuations.

Such a cyclic pattern would be established if more than an average proportion of mature females conceived during one exceptionally 'good year'. The long gestation period (20–22 months) and the subsequent period of lactational anoestrus would make a large proportion unavailable for conception for two to three years following the good year. This would result in a high birth rate in one year followed by lower rates in the next two years. Douglas-Hamilton (1972) found that the number of calves born to a sample of 98 identified female elephants varied from 8 to 34 annually. Variation in the annual mortality of young calves may also distort the true birth rate. A drought may affect calves proportionately more than older animals.

For the Asian elephant also there is evidence of such cycles in annual recruitment. The sample of 88 elephants captured during 1968 in the Mysore Kheddah included a number of entire family groups. The age structure of this sample is shown in Fig. 11.2. Whereas there were only two and three elephants in the 1–2 and 2–3 age classes respectively, the 0–1 age class had 13 calves, most of them below six months. Further, five more calves were born within ten weeks of capture. Thus, 18 calves were born to 30 adult cows within a span of about one year, a very high birth rate which would be totally

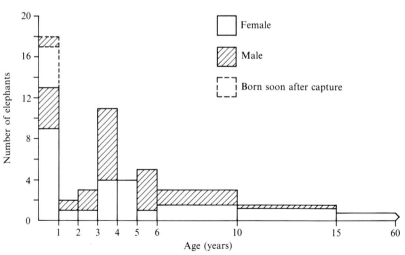

Fig. 11.2. Age structure of elephants ($n = 88$) captured in the Mysore Kheddah of 1968.

Fertility

misleading if only one year's data were to be considered. The age structure strongly suggests a cycle in recruitment with a period of three years.

Two methods were used to estimate the fertility of the study population. One was based on the annual birth rate and the other based on the difference in age between two offspring of a mature female. The birth rates during 1981, 1982 and 1983 were considered. It was not possible to follow strictly the method of Douglas-Hamilton (1972), who considered the number of births in each year to a set of 98 identified cows. This was because many of the identified females in the study population were not recorded every year. Instead, the number of calves below one year of age was related to the number of distinctly identified mature females (Table 11.3).

If mature females are taken as those above 17.5 years (the mean age at first calving), then there was a minimum of 23 calves per 101 females during the three years, indicating a fertility rate of 0.228 births per mature female per year or a mean calving interval of 4.4 years. If each year is given equal weighting, the fertility rate is 0.21 and the calving interval 4.8 years.

A clarification must be made here about the extremely low number of calves recorded in 1983. This could have been due to two reasons. First, the relatively high birth rate during 1981 and 1982 would have naturally made a large proportion of cows unavailable for conception. The year 1983 could thus lie in the trough of a cyclic recruitment pattern. Second, the low rainfall from June 1982 to May 1983 could have resulted in high calf mortality. Only one young calf was, however, found dead in the study area during January–June 1983. Even considering that dead calves could have escaped notice in forested habitats, it is unlikely that calf mortality alone could have caused the trough in the 0–1 age category. It is more likely that the 1983 sample represents a natural year of low number of births.

In African elephants the rate of conception in a year has been shown to be positively correlated with the annual rainfall (Douglas-Hamilton 1972) or apparently unrelated to rainfall (Hanks 1972a). In this study there was no clear evidence that the birth rate in a year was related to the annual rainfall two years prior to it, as the figures in Table 11.4 illustrate.

Table 11.3. *Annual birth rate of the elephant population*

	1981	1982	1983
Number of registered females above 17.5 years	36	39	26
Number of calves below 1 year	8	13	2
Birth rate (number of calves per mature female)	0.22	0.33	0.08

Table 11.4. *Relationship between rainfall and birth rate*

Station	Total annual rainfall (cm)		
	1979	1980	1981
Attikan–BRT Betta	178	118	196
Hasanur	208	76	94
Birth rate (number of calves per mature female)	0.22	0.33	0.08
	(1981)	(1982)	(1983)

Another approach to the estimation of the mean calving interval was to age the last two offspring of identified female elephants. The difference in age between the two offspring is taken as the calving interval. Only those offspring below 9 years were considered. Fourteen such mother–offspring units gave a mean calving interval of 4.6 years (s.d. = 1.07). The shortest calving interval recorded was about 3 years and the longest 6.5 years.

Since the mean calving intervals calculated by the two methods closely agree, a figure of 4.6–4.8 years can be taken as reasonably accurate for the study population.

11.3 Mortality

Elephants may die of natural causes or they may be directly killed by people. Causes of natural death include disorders of the gastrointestinal tract, pulmonary and cardiovascular disease, miscarriage, starvation in old elephants due to wearing out of the last set of molar teeth, accidental fall from a steep slope and injuries from fights between bulls. Bacterial diseases such as anthrax have also been occasionally recorded. The frequency distribution of natural mortality during different months is presented in Table 11.5, based on a sample of 194 elephant deaths between 1978 and 1983. There was no increased mortality during any particular season including the dry spell.

In southern India about 18% of all female elephant deaths and 70% of male deaths were directly due to man. The female deaths were in defence of crops; the killing of male elephants was either to prevent crop damage or to steal the tusks. This aspect has already been elaborated in Chapter 10.

This section will be concerned not with details about the various causes of death in elephants but with the actual numbers dying and the age-specific mortality rates. It is rather difficult to estimate mortality rates in wild mammalian populations. The applications and the limitations of various methods have been discussed by Caughley (1977). Although it would be futile to seek the exact mortality rates for each age class, it is still profitable to

Mortality

deduce the approximate age-specific rates by different methods. Three approaches were taken to study mortality patterns: the number of elephants found dead in the study area in relation to total population size, a life-table for female elephants constructed from a sample of dead elephants, and the mortality rate indicated by the shrinking of successive age classes of living elephants.

11.3.1 *Numbers found dead in the field*

During 1981 and 1982 the number of elephants found dead in the main study area included 12 females and 20 males. These figures thus represent the minimum number of deaths. It is likely that a high proportion of carcasses were discovered because the area is traversed daily by livestock graziers, who are usually the first to notice and report elephant deaths.

In the main study area, the average number of elephants was estimated to be 333 females and 157 males. A crude measure of the annual mortality rate (assuming a stationary population) is 1.8% for females and 6.4% for males. If only elephants older than 5 years are considered, the annual rates work out to 1.7% (4.5/265) for females and 11.8% (9/76) for males. The true figures may be higher. In particular, the carcasses of young animals tend to go unnoticed and additional deaths in the 0–5 year age class usually have to be inserted (Laws 1969).

This simplistic approach does not reveal anything about the age-specific mortality. If the standing age distribution with population sizes and the number of deaths that occur in each age class during a time interval are known, it is possible to construct a 'current life table' with age-specific death rates (Pielou 1977). This again assumes that the age distribution, usually taken at the mid-point of the time interval, does not change during the time interval.

Table 11.5. *Frequency distribution of natural mortality during 1978–83*

Month	J	F	M	A	M	J	J	A	S	O	N	D
Number of natural deaths	16	15	18	17	16	15	14	20	18	11	19	15

The total sample size is 194 natural elephant deaths in the Nilgiris – Eastern Ghats population of southern India.

Statistical test:
The distribution of deaths was tested for statistical significance by the *G*-test, to see whether the risk of mortality was higher during any month. The frequency distribution does not deviate from an even distribution ($G = 4.3$, d.f. $= 11, p > 0.10$).

This exercise was carried out for the female segment of the study population, but it is not worthwhile to present the data. With a small sample of 12 deaths in two years out of 333 elephants it would be unrealistic to expect any clear age-specific pattern. On the whole, the death rate in any age-class did not vary much from the mean rate of 1.8% per annum.

For the male segment, with a sample of 20 deaths it was more useful to compile a current life-table and get an idea of the age-specific death rates. The life-table derived in this fashion is given in Table 11.6. For male elephants above 10 years, the mortality rate averages 15% per annum owing to the high incidence of poaching. The apparently low mortality of juveniles is partly due to incomplete representation of such carcasses. Given all these limitations, however, both male and female juvenile mortality seems to have been very low in the study area during 1981–82.

Of the 20 male deaths only five were natural deaths and the rest due to man. In the absence of unnatural death the number of natural deaths would be higher because a certain proportion of the elephants killed by man would otherwise have succumbed to natural causes. One difficulty in getting a measure of male mortality is the fluctuation in the incidence of poaching within a particular region. Whereas 1981 and 1982 were years of relatively high poaching, the rate came down in 1983. It was tentatively estimated that the annual mortality rate of males above 5 years could have been around 10% during the 1970s.

Table 11.6. *Current life-table for male elephants*

Age class (years)	Total population at midpoint of time interval Q_x	Number of deaths in the age class per year D_x	Age-specific death rate (per cent) $M_x = D_x/Q_x$
0–5	81	1.0	1.2
5–10	29	2.0	6.9
10–15	12	2.5	20.8
15–20	6	0.5	8.3
20–30	10	1.5	15.0
30–70	18	2.5	13.9

Age frequencies Q_x have been smoothed.
D_x values are based on the numbers dying during 1981–82.

11.3.2 Life-table from ages at death

If a population has reached a stable age distribution and its instantaneous growth rate, r, equals zero, then a sample of dead animals picked up from the field can be treated as a group of individuals or cohort born at the same time and dying at different ages. By knowing the ages at death and the proportion dying in each age class, a composite age-specific life table can be constructed. If r is not zero, but its value is known, a correction factor has to be introduced (Caughley 1977). Most life-tables for mammals have been constructed in this fashion. Although they are a useful exercise in understanding mortality patterns, few such life-tables are satisfactory in estimating absolute age-specific mortality rates in view of the many assumptions made.

For the study population the age distribution of male elephants was unstable owing to irregularities in age-specific mortality from poaching. A life-table constructed in this fashion would certainly be invalid. The distribution of ages at death for males are presented merely for the record in Table 11.7, based on a sample of 200 elephants which died during 1977–83 in the Nilgiri – Eastern Ghats population of southern India. This sample included 105 natural deaths and 95 deaths due to man. The latter category is under-represented because the ratio of natural to unnatural deaths is around 35:65 (Chapter 10). The two categories have been corrected by this ratio and the combined age distribution at death given as a percentage of total male deaths.

Table 11.7. *Distribution of ages at death for male elephants*

Age class (years)	Distribution of natural deaths	Distribution of deaths caused by man	Weighted distribution of deaths* (per cent)
0–5	34	0	11.3
5–10	13	17	15.9
10–15	14	17	16.3
15–20	5	18	14.0
20–30	14	19	17.7
30–70	25	24	24.7
Total	105	95	100.0

* The weighted distribution is based on a ratio of 35 natural deaths: 65 deaths caused by man (see Chapter 10).

For the female segment it is more reasonable to assume an age distribution close to the stable one and a value of r also close to zero. A sample of 90 elephants which died during 1977–83 in the Nilgiri – Eastern Ghats region provided the raw data. Elephants were grouped into age classes. It was not possible to assign age classes beyond 20 years from body measurements. The life-table and inferred mortality rates are shown in Table 11.8.

This analysis indicates a female mortality rate of nearly 5% per annum in the 0–5 age class and 2–3% between 5 and 15 years. The higher mortality in the 15–20 age class could be due to the risks of the first pregnancy. There was evidence that a number of these deaths were of pregnant elephants, some of which miscarried. Above 20 years it was possible to get only an approximate rate. The relatively high rate of over 7% could be due to an increased mortality after 40 years. Other evidence indicates that female elephants have a low mortality of only 3% or so per annum between 20 and 40 years.

A mention must be made of the difference in mortality rate of males and females in the 0–5 age class. In a sample of 290 dead elephants that were sexed and aged, there were only 19 females in the 0–5 age class as against 34 males in the same class. Since the sex ratio at birth is equal, the higher number of male deaths (by a factor of 1.8) could only be due to a male mortality rate twice as high as the female mortality rate between birth and 5 years.

11.3.3 Instantaneous mortality rate from the standing age distribution

Laws *et al.* (1975) demonstrated that the mortality rate for much of the lifetime (excepting juvenile and old age) is nearly constant in a number of

Table 11.8. *Life-table for female elephants*

Age class (years)	Mortality d_x	Survival l_x	$\ln(l_x)$	Slope $-z$	Survival rate e^{-z}	Mortality rate per year $1 - e^{-z}$
0		100	4.605			
0–5	21.1	78.9	4.368	−0.047	0.954	0.046
5–10	8.9	70.0	4.248	−0.024	0.976	0.024
10–15	6.7	63.3	4.148	−0.020	0.980	0.020
15–20	17.7	45.6	3.820	−0.066	0.937	0.063
20–70		1.0*	0.000	−0.076	0.926	0.074

Survivorship based on a sample of 90 dead elephants.
* It has been assumed for convenience that 1% of females survive until age 70 years.

Mortality

African elephant populations. When the natural logarithm of the age class abundance of living elephants is plotted against age, it is linear over the range 5–50 years. The instantaneous mortality rate is estimated from the slope ($-z$) of the fitted regression. A stable age distribution and an r value of zero have to be assumed.

This approach would be invalid if the inherent assumptions did not hold good. For instance, if a sudden increase or decrease in mortality affected all the age classes equally, the resulting age distribution would be the same as the earlier age distribution. The slope of the regression would also be the same for both these distributions, yet their mortality rates would be quite different (Caughley 1974). This exercise was attempted only because there was no evidence from the available records for any sudden increase or decrease in female mortality in the study area and elsewhere. Such drastic changes in mortality may be more common in African elephant populations (Corfield 1973; see also Chapter 6).

Regression lines were fitted to the different segments of the age structure (Fig. 11.3 and Table 11.9). Correlation coefficients in all cases were high. Between 5 and 40 years, the z value of 0.033 corresponds to an annual survival rate (e^{-z}) of 0.968 or annual mortality of 0.032 (3.2%). The logarithmic plot and the values of z over different segments of age also suggest an increased mortality of about 7% at higher ages.

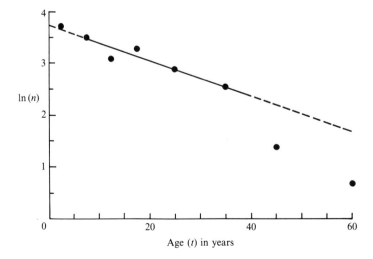

Fig. 11.3. Instantaneous mortality rate in female elephants. The line represents $\ln(n) = 0.033\, t + 3.73$, where n is abundance and t is age in years, for elephants aged 5–40 years.

Table 11.9. *Instantaneous mortality rate for different age segments in female elephants*

Age class (years)	Instantaneous mortality rate z	Correlation coefficient r
0–50	0.048	−0.96
5–40	0.033	−0.93
10–50	0.054	−0.92
30–70	0.072	−0.97

11.4 Population modelling

At the first instance, a fixed fertility schedule was used along with the nine possible combinations of three mortality schedules each for male and female elephants. Later, in specific cases the fertility rates were varied to see their effect on the growth rate of the population.

The assumptions made in the simulations were as follows:

Age at first calving: 17.5 years
Age of menopause: 60 years
Mean calving interval: 4.7 years

Separate sets of age-specific mortality schedules for male and female elephants were used (Table 11.10). Three mortality patterns (low, medium and high) were considered for each sex. However, a mortality pattern for males in one category was taken to be higher than the mortality pattern for females in the corresponding category. There was a certain rationale behind the selection of these mortality rates. On the whole, the low mortality schedules represent the minimum levels expected in the population, the medium mortality schedules the suspected average levels operating in the long term, and the high mortality schedules the levels that could be easily reached during certain adverse years.

For the present it was felt that invoking a density-dependent negative feedback on the population growth rate was unnecessary. In large mammals there is evidence that density-dependent change is non-linear and begins to operate only when the population is close to the carrying capacity (Fowler 1981). For the study area there was no concrete evidence that the elephant population had reached the level where such a feedback could have begun to operate (see Chapter 6). Thus, only fixed values of the parameters were used. Simulations were run for a period of 50 years by which time the stable age distribution was reached.

Table 11.10. *Mortality schedules used in the simulations*

Age class	Years	Mortality rate[a]		
		Low	Medium	High
Females				
Infant	0–1	5	10	15
Juvenile	1–5	4	8	12
Sub-adult	5–15	2	3	4
Adult	15–20	3	5	7
	20–40	2	3	4
	40–50	4	6	10
	50+	10	12	15
Males				
Infant	0–1	8	15	20
Juvenile	1–5	6	12	16
Sub-adult	5–15	6	10	15
Adult	15–20	6	10	15
	20–40	6	10	15
	40–50	8	10	15
	50+	10	10	15

[a] Rates are given as percentages per year.

11.4.1 Trends in population growth

Since the rates of population growth are very small in elephant populations, these have been expressed as percentage per year, rather than as r or e^r, for easier understanding.

When the fertility is kept constant, the population growth rate is determined primarily by the female mortality schedule. The growth rate can be considered over two time periods: the immediate future and in the long term after the age distribution has stabilized. Average growth rates have been given for 1–5 years and 46–50 years of constant fertility and mortality (Table 11.11; Fig. 11.4).

With the current fertility schedule and a low female mortality, the growth rate that could be expected in the long term is about 1.6% per annum. A high female mortality results in a population decline at a rate of 2% per annum. With a medium female mortality, the growth rate is very close to zero (−0.2%), indicating a stable population size. However, the parameters need to change within only a relatively small range to produce either a declining or an increasing population.

Because juvenile mortality seemed low in both sexes from available evidence, two more simulations were run. These incorporated a low juvenile mortality for both sexes, a medium sub-adult/adult female mortality, and either a medium or a high sub-adult/adult male mortality. In both cases the growth rate stabilized at 0.5% per annum. Since either a medium or a low juvenile mortality may be operating in the population, it was concluded that the population was stable or increasing at a slow rate.

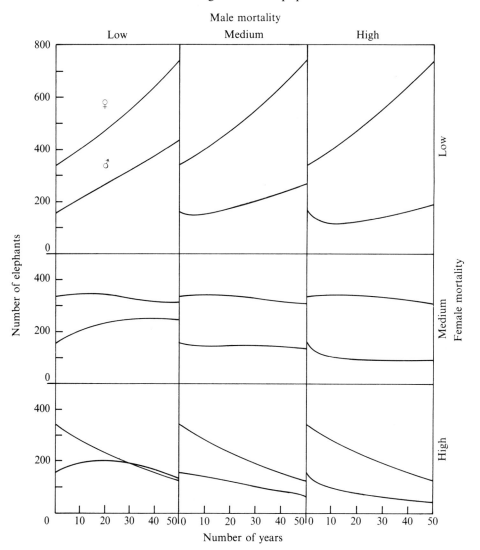

Fig. 11.4. Simulated trends in population numbers with different mortality schedules. The upper line refers to the female segment and the lower line to the male segment of the population.

Table 11.11. *Population growth rate with different mortality schedules*

	Male mortality			Female mortality
	Low	Medium	High	
(1)	+2.32	+1.06	+0.21	Low
(2)	+1.55	+1.55	+1.55	
(1)	+1.05	−0.27	−1.36	Medium
(2)	−0.16	−0.20	−0.20	
(1)	−0.16	−1.58	−2.72	High
(2)	−1.84	−1.97	−1.99	

The values are the mean annual rates of growth in percentage during (1) 1–5 years and (2) 46–50 years.

What is the maximum growth rate that could be expected in an Asian elephant population? It is unlikely that mortality could decrease below the low female – low male mortality schedule. The mean calving interval of 4.7 years also compares favourably with the most productive African elephant populations. These parameters were kept constant and the mean age at first calving was lowered from 17.5 to 12.5 years. Even then the growth rate achieved in the long term is only 2.4% per year.

11.4.2 Trends in age distribution

A positive growth rate is usually associated with a predomination of the younger age classes and a negative growth rate with a shift towards the older age classes. Often the population growth rate has been sought merely from the standing age distribution, an approach strongly criticized by Caughley (1974, 1977).

These simulations also brought out the unreliability of interpreting population growth from age distributions (Tables 11.12 and 11.13). When the age distributions are taken as a proportion of the total population, a shift towards the adult age classes occurs for one sex if the mortality rate for that sex is decreased and for the opposite sex increased. The total proportion of adults (males plus females) increases notably, from 42 to 51% of the population, only under one condition: when male mortality is kept low and female mortality increased to a high level (growth rate is negative). This, of course, means that mortality rates for the two sexes are nearly equal.

Population dynamics

Table 11.12. *Trends in the percentage of calves in the population*

Male mortality			Female mortality
Low	Medium	High	
5.74	6.74	7.28	Low
5.42	6.72	7.44	Medium
4.88	6.58	7.54	High

The percentage values are given after 50 years of constant fertility and mortality and refer to calves younger than 1 year, beginning with a value of 7.1%.

Table 11.13. *Trends in the percentage of adult elephants in the population*

	Male mortality			Female mortality
	Low	Medium	High	
M% F%	11.5 31.5	4.2 36.9	1.4 39.9	Low
	43.0	41.1	41.3	
M% F%	17.1 29.0	6.3 36.0	2.1 39.8	Medium
	46.1	42.3	41.9	
M% F%	25.2 25.5	9.6 34.4	3.0 39.5	High
	50.7	44.0	42.5	

The percentage values are given after 50 years of constant fertility and mortality, beginning with 6.5% adult male and 35.5% adult female in the population.

The proportion of calves below one year in the total population is also no clear indication of whether the growth rate is negative or positive. Thus, the percentage of calves (7.5%) in a population which is clearly declining under a high male – high female mortality schedule is higher than the percentage of calves (5.7%) in a population that is increasing due to a low male – low female mortality. It is only when the differences between male and female mortality rates are eliminated or kept relatively low that a decline in the growth rate will also show up in a decline in the proportion of calves in the population. This is being stated specifically because there is often a misconception that the proportion of calves in the population indicates the trend in growth rate. For instance, Field (1971) appeared puzzled that a population could increase in spite of a decline in the proportion of calves from 8.5% in 1963 to 5.6% in 1968–69 in the Queen Elizabeth National Park, Uganda.

The relationship between age distribution and the population growth rate is a complex one dependent on the differential mortality between the sexes and also on the fertility schedule. Consider two populations: population A which increases primarily due to an increase in fertility, while population B increases due to a decline in mortality. In population A the increased recruitment will shift the age distribution towards the younger age classes. In population B, if the change in mortality is proportional for all age classes of both sexes, there will be no change in the age distribution.

In these simulations the fertility was kept constant and the mortality rates varied; hence the ambiguity in trends in age distribution with changing growth rates. A naive interpretation of age ratios, without other information on fertility and mortality rates for the sexes, can be very misleading, a point well brought out by Caughley (1974). Inferences of relative trends in growth rates between two populations cannot usually be made merely by comparing their age distributions unless other information is available.

In certain cases the age structure of an elephant population may indicate the sign of r, though not its absolute value. For example, the abnormal depletion of elephants below 20 years in Murchison Falls Park, Uganda, was clearly an indication of declining fertility and a negative growth rate (Laws *et al.* 1975).

11.4.3 *Trends in the adult sex ratio*

The sex ratio at stable age distribution will depend on the magnitude of the difference in mortality rates of male and female elephants (assuming an equal ratio at birth). The simulations began with a ratio of 1 adult male:5.4 adult females. With a low male mortality the disparity narrowed, but for a high male mortality it widened considerably (Fig. 11.5).

194 *Population dynamics*

For a high male – medium female mortality, which was easily possible due to male deaths from poaching, the ratio ultimately reached about 1 male:20 females. For the medium male – medium female mortality the ratio stabilized at 1:5.7, which is very close to the initial ratio. Two simulations incorporating a low juvenile mortality were run to see the trends with a medium or a high male mortality, keeping the adult/sub-adult female mortality at a medium level. With a medium male mortality of about 10% in the sub-adult and adult classes the adult sex ratio stabilized at 1 male:4.9 females. If male mortality increased to 15% due to poaching, the adult sex ratio reached 1:12.7.

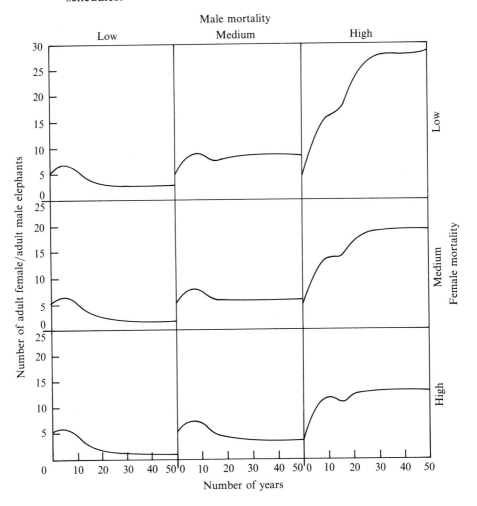

Fig. 11.5. Simulated trends in the adult sex ratio with different mortality schedules.

These results strongly suggest that the medium mortality schedules for both sexes used in the simulation could have been operating in the past in the population. With a spurt in poaching this could further widen to between 1:10 and 1:20 if hunting levels are not reduced.

Another important trend noticed was that even if male mortality were to be reduced to a low level, the adult sex ratio would further widen over a short term of about 5 years before narrowing down. This is due to a depletion in the 7–10 and 10–15 age classes in the current age structure.

11.4.4 Possible effect of a disparate adult sex ratio on fertility

For the Asian elephant a factor that must be seriously considered is the possible influence of the disparate adult sex ratio on the fertility of the population. In most mammalian populations the adult sex ratio is biased in favour of females owing to a higher natural mortality rate in males. In a polygynous society there is usually a 'surplus' of males, since one male can mate with numerous females. During any particular period, a certain proportion of adult cows would also be pregnant or in lactational anoestrus, and hence unavailable for conception. The operational adult sex ratio would therefore not be as disparate as the observed sex ratio in the population. Thus, a normally disparate sex ratio can still ensure that most if not all the mature females available for breeding during a season are mated.

Hunting of the male elephant further widens this disparity in the sex ratio. It is obvious that this cannot widen indefinitely without causing a decline in the fertility. At some ratio of adult male to adult female there will be too few males to ensure that all the available females are successfully mated, resulting in a lower rate of conception and a longer inter-calving interval. In theory this lowered fertility could reduce the rate of population growth.

The crucial question is at what ratio of adult male to female this factor will begin to operate. Will it begin at 1:2 or 1:10 or 1:100? To derive an empirical answer the mean calving interval will have to be determined in populations with varying sex ratios. Even then this relationship may not be clear since conception, and hence calving interval, will also be influenced by factors other than the sex ratio, such as rainfall, nutrition level in females, etc. To model this system theoretically, information on many aspects of reproductive behaviour in elephants is needed, not all of which is available.

In a population having a pronounced breeding season it is more likely that an available female will not conceive if males are a scarce resource. If breeding activity is spread throughout the year, the shortage of males is less likely to be important. In the study area, with rainfall spread over a period of 8 months, there was no evidence for seasonal breeding.

Population dynamics

Table 11.14. *Population growth rate with different calving intervals*

Mortality	Adult sex ratio M:F	Population growth rate with calving interval		
		4.7 years	5.5 years	7.7 years
Male high, female low	1:28.7	+1.55%	+1.08%	0
Male high, female medium	1:19.4	−0.20%	−0.66%	—

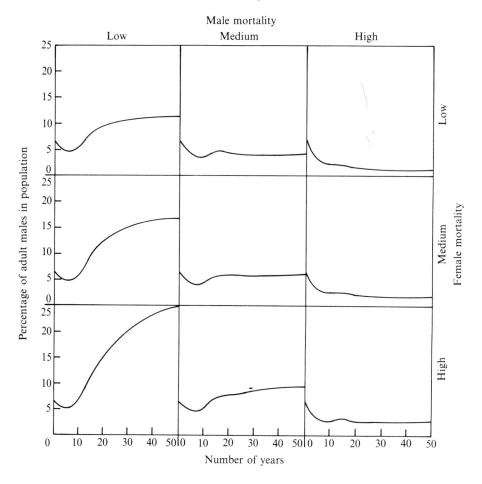

Fig. 11.6. Simulated trends in the proportion of adult males in the population with different mortality schedules.

Although no objective criteria for deducing a decline in fertility were available, the effect of a higher calving interval on the population growth rate for instances of a very disparate ratio was simulated (Table 11.14).

When the mean calving interval was increased from 4.7 to 5.5 years, the growth rate changed by less than 0.5% per year. Another simulation showed that with a high male – low female mortality the growth rate became zero only when the mean calving interval reached 7.7 years. These results illustrate that as long as female mortality is fairly low the population could still have the capacity to increase or remain stable in spite of a decrease in fertility due to higher male mortality.

11.4.5 A test of the validity of the model: predicted and observed trends in the proportion of adult males

The model predicted a decline in the proportion of adult males in the population over a period of 5–7 years, even if male mortality were to be reduced from the prevailing medium–high schedules to a low schedule, due to a depletion in the 7–10 and 10–15 year age classes (Fig. 11.6).

With a medium–high mortality rate indicated by the data, the predicted trends were a decline in the proportion of adult males from an initial 6.5% to around 4% during the fifth year. The same population was sampled over the larger Eastern Ghats – Nilgiri region during 1987. It was possible to get only an approximate classification into juvenile, sub-adult and adult classes. Out of 1188 elephants that were sexed and classified during January–October 1987, the adult males enumerated were 46 or 3.9% of the population. Assuming that a few tuskless males may have been overlooked, the proportion of adult males had come down to 4.0–4.5%, which corresponds to the predicted level (Table 11.15). This first test of the model indicates that the methods used in estimating various demographic parameters are basically robust. The model can be profitably used in future monitoring of population trends.

11.5 Demographic condition of the population

The condition of a population can be characterized by its 'demographic vigour' (Caughley 1977) and its 'physiological condition' (reviewed by Hanks 1981). Caughley (1977) proposed that the rate of increase, which he symbolized as r^s, implied by the prevailing schedules of fertility and mortality, provides a measure of the demographic vigour of the population. Changes in the environment would result in a change in the value of r^s. Hanks (1981) rightly pointed out that the value of r^s itself does not tell anything about the recuperative powers of a population. Thus, one population may

Table 11.15. *Predicted trends in the proportion of adult males in the population*

	Number of years from the base year (1983)											
	1	2	3	4	5	6	7	8	9	10	25	50
Proportion of adult males with												
(a) Medium male and medium female mortality	6.1	5.7	5.2	4.8	4.4	4.6	4.7	4.7	4.7	4.7	6.0	6.3
(b) High male and medium female mortality	5.9	5.2	4.6	4.0	3.5	3.5	3.3	3.2	3.0	2.8	2.1	2.1

The proportion of adult males during 1987 was between 4% and 4.5% of the population (see text).

Demographic condition

show a high rate of increase but recover slowly after a sudden crash, owing to environmental stress, while another population may be stable but recover rapidly after a decline. He favoured a measure of resilience based on the physiological condition of the animals.

No single demographic or physiological parameter can accurately assess the condition of a population and predict future trends. There is evidence that the dynamics of large-mammal populations are influenced by stochastic environmental perturbations or long-term population cycles which are little understood (Wu & Botkin 1980; Croze *et al.* 1981; see also papers in Sinclair & Norton-Griffiths (1979)).

Nevertheless, if management goals require an assessment of merely the present condition and possible short-term future trends, both demographic and physiological parameters can be used profitably (Hanks 1981). In this study it was not possible to measure any index of physiological condition; hence, only the demographic parameters of the study population are discussed.

A comparison of demographic parameters in various Asian and African elephant population is given in Tables 11.16 and 11.17. The age of sexual maturity in females in the southern Indian population is high compared with that in African elephant populations known to have positive growth rates. This need not, however, be interpreted as a loss of demographic vigour. The

Table 11.16. *Comparison of fertility parameters in different elephant populations*

Region	Year	Age of sexual maturity in females (years)	Mean calving interval (years)	Status of population
1. Satyamangalam–Chamarajanagar, Southern India	1981–83	15.0	4.7	—
2. Mkomazi (East), Tanzania	1968	12.2	2.9	increasing
3. Mkomazi (Central), Tanzania	1969	12.2	4.2	increasing
4. Tsavo NP, Kenya	1966	11.7	6.8	—
5. Murchison (North), Uganda	1966	16.3	9.1	—
6. Murchison (South), Uganda	1967	17.8	5.6	declining
7. Budongo (Central), Uganda	1966	22.8	7.7	—
8. Luangwa valley, Zambia	1967–69	14.0	3.5–4.0	—
9. Lake Manyara, Tanzania	1966–70	11.0	3.9–4.6	increasing
10. Kruger NP, South Africa	1970–74	12.0	4.5	increasing
11. Kasungu NP, Malawi	1978	—	3.9	increasing

Sources: 1, this study; 2–7, Laws *et al.* (1975); 8, Hanks (1972a); 9, Douglas-Hamilton (1972); 10, Smuts (1977); 11, Jachmann (1980).

age at first calving may vary across different climatic regimes. The African bush elephant in the unpredictable semi-arid zones could be expected to have a lower age of sexual maturity than an Asian elephant population in a relatively more stable environment (see Chapter 6).

The mean calving interval of 4.7 years recorded in the study area compares well with the most productive African populations at Kruger, Mkomazi Central and Lake Manyara. This indicates that mature females have a reasonably high fertility.

Hanks (1981) suggested that subnormal foetal development and an increase in juvenile mortality are definite indications of declining demographic vigour. In Tsavo National Park, an annual mortality rate of 36% during the first year and 10% for 1–5 years was estimated even prior to the population crash of 1970–71 (Laws 1969; Corfield 1973). In Lake Manyara the calf mortality rate was 10% during the first year (Douglas-Hamilton 1972). For the study area and adjoining regions the available data indicate that juvenile mortality was at relatively low levels. From birth to 5 years the annual mortality was only 4–5% in female and 8–9% in male elephants. Female mortality from 5 to 40 years was also at a low 2–3% annually and, as shown by the simulations, this could be an important factor in maintaining a positive or zero growth rate in the population.

A population in poor physiological condition could be expected to show a higher mortality during the dry season, especially when there is a drought. Since the data show no significant increase in natural mortality during the dry period compared with the wet season it is reasonable to assume that the

Table 11.17. *Comparison of age and sex class proportions in different elephant populations*

Region	% adults in the population		% sub-adults in the population		% calves in	
	Male	Female	Male	Female	Total population	Family herds
1. Chamarajanagar–Satyamangalam, Southern India	7.0	35.4	21.5	29.0	7.1	7.7
2. Gal Oya, Sri Lanka	22.7	44.8	?	?	6.1	7.3
3. Lahugala, Sri Lanka	16.1	38.7	?	?	8.9	10.6
4. Yala NP, Sri Lanka	15.7	29.2	18.6	26.5	10.0	12.0
5. Kruger NP, South Africa	19.6	21.5	25.9	26.2	6.8	8.5
6. Lake Manyara, Tanzania	10.7	27.2	?	?	6.8	7.6

For convenience adults are taken as those above 15 years, sub-adults as 1–15 years and calves as below 1 year.
Sources: 1, this study; 2 and 3, McKay (1973); 4, Kurt (1974); 5, Smuts (1977); 6, Douglas-Hamilton (1972).

elephant population is in good physiological condition. The population was also extremely resistant to the droughts of 1982–83 and 1986–87.

Between 1977 and 1982 the bacterial disease anthrax was not recorded in elephants in Karnataka and Tamilnadu states. During 1983–86 at least 11 elephants died of anthrax in Bandipur and Mudumalai. This disease may have been transmitted by cattle. Although there is no historical record of an epidemic outbreak among elephants in southern India, the appearance of anthrax warrants a close watch on the populations.

The high mortality in male elephants due to poaching is of some concern. The proportion of adult males in the southern Indian populations is the lowest for any known Asian or African elephant population. Apart from the possible effect of a highly disparate sex ratio on the fertility of the population as mentioned earlier, the genetic consequences of too few breeding males will also have to be considered. This would include inbreeding depression and a loss of genetic variation through drift, especially in small isolated populations. These aspects are discussed in more detail in Chapter 12.

12

Conservation and management

It is being increasingly realized that conservation efforts should expand from a focus on single species, typically large animals such as crocodiles, cranes or the tiger, to the overall biological diversity of the earth. The Asian elephant today ranges over a wide spectrum of vegetation types; hence, its conservation can be integrated with the conservation of biological diversity. Elephants make a significant impact on the ecosystem. Since a major portion of their habitat is hill forest, the management of these areas as 'elephant ranges' would influence the quality of watersheds, longevity of dams and land-use in the plains (McNeely 1978). Elephants still have an important non-polluting, non-destructive role in the logging industries of Burma, Thailand and India.

The elephant can survive in a developing country, having a large human population living at subsistence levels, only if it is accepted by the people who share its habitat and resources. Any grandiose plan for conservation without adequate provision for human interests is doomed to fail. Conservation in developing countries often has to be a compromise between idealism and reality. This chapter summarizes the scientific principles of elephant management and explores means of minimizing the conflict between elephants and people. Conservation efforts should ideally tackle both these issues concurrently.

12.1   Conservation of the elephant
12.1.1   *Minimum viable population size*

The smaller the population size the more vulnerable it is to extinction due to demographic, environmental and genetic stochasticity (Shaffer 1981; Gilpin & Soulé 1986). A small population may become extinct through chance fluctuations in fertility and mortality of individuals or a peculiar age-structure. For instance, Shaffer (1981) estimated that a grizzly bear population of less than 30–70 individuals, depending on demographic characteristics, has less than 95% chance of surviving for 100 years. Environmental stochasticity may result from a disease epidemic, or adverse climate such as a severe drought, which could wipe out a small population.

Another consideration is the genetic viability of a population. In small populations the gene frequencies change randomly from one generation to another with a fixation or loss of alleles. This process is known as genetic drift. Ultimately, the loss of alleles leads to an increase in homozygosity or genetic uniformity. The central question in conservation genetics is the relationship between genetic variation and the fitness of the individual and the species. Increased homozygosity has been associated with a lower fitness (Frankel & Soulé 1981; Allendorf & Leary 1986).

In the short-term, the most serious consequence of a small population is inbreeding depression (Ralls, Brugger & Ballou 1979; O'Brien et al. 1985). The immediate effect of intensive inbreeding is a loss of fitness through lowered fertility, higher juvenile mortality, depressed growth, etc. Based on experiences with breeding of domestic mammals, it has been suggested that a minimum of 50 effective breeding individuals is needed to keep inbreeding depression to a negligible level of below 1% inbreeding per generation (Franklin 1980).

The issue of long-term fitness of a population, in terms of evolutionary potential, is still rather speculative. There have been attempts to theoretically derive the minimum population size above which the effects of genetic drift can be countered through natural selection or by gain from mutation (Franklin 1980; Frankel & Soulé 1981). From both angles a figure of about 500 effective breeding individuals has been derived. Populations maintaining this effective size can be expected to remain viable from an evolutionary viewpoint.

A clarification has to be made here regarding the 'effective population size'. The census figure N for a species constitutes the genetically effective population size N_e only under idealized conditions such as an equal sex ratio of breeding individuals, an equal number of progeny per mating pair per generation and no fluctuation in population size. Of these the most important issue for elephants is the sex ratio. Since male elephants suffer a higher mortality than female elephants from natural causes and poaching, the sex ratio of adults is usually unequal. At best it may be 1 male : 2 females; at worst it may go up to 1 : 20 or even more disparate as in parts of southern India. The more unequal the sex ratio, the higher is the rate of genetic drift. The effective population size N_e is equal to $4N_m N_f/(N_m + N_f)$, where N_m and N_f are the numbers of breeding males and females, respectively. The value of N_e decreases with an increasingly disparate sex ratio.

What are the implications of conservation genetics theory for the elephant populations in Asia? In southern India, the largest population inhabiting the Nagarhole – Nilgiris – Eastern Ghats region consists of 3600–4300 elephants. Of these 7% were adult males and 35% were adult females during 1981–83. Taking the lower population estimate there were 252 males and 1260 females capable of breeding. Using the formula given above this translates into an effective population size of 840 breeding individuals, a comfortable level to counter the loss of genetic variation through drift. By 1987 the proportion of adult males had been reduced by poaching to 4.5% of the population. The effective population size has thus come down to about 574 breeding individuals (assuming 162 males and 1260 females). This example shows how changes in the sex ratio alone, without any significant drop in total numbers, can lower the genetic fitness of a population. Large populations exist in other regions of southern India such as the Anamalais and the Periyar plateau. Rampant poaching, especially in the latter area, has drastically reduced the number of male elephants, resulting in a highly disparate sex ratio and a reduction of the effective population size to, perhaps, below 500 individuals.

In northeastern India there are three large populations. These are found in the Arunachal Pradesh – north Assam region, the Kaziranga–Naga hills, and the Garo hills – Khasi hills of Meghalaya. Since 50% of the male elephants are tuskless, the sex ratio is not likely to become as highly skewed, as in southern India, due to poaching, although the frequency of tuskless males could increase in the population.

Other known elephant populations in Asia which may maintain effective population sizes above 500 are those in the Irrawady–Chindwin valleys and Northern hill ranges of Burma, in southeastern Sri Lanka and in Sabah.

Since most Asian elephant populations are small and isolated, the goal should be to maintain an effective population size of at least 50 breeding individuals to counter inbreeding depression in the short term. A population having 20 adult males and 40 adult females satisfies this criterion. This translates into a total population of 125–150 elephants. If the sex ratio is more unequal than 1:2 these figures will correspondingly increase.

If an elephant population has been recently split such that the resulting populations have effective sizes of less than 50 individuals, these may be managed through an artificial exchange of individuals. It would be most convenient to translocate adult male elephants from one population to another; this would mimic the natural male dispersal from family groups. The migration of even one elephant per generation may help to maintain genetic variation and counter inbreeding. Exchange between populations inhabiting very different environments is not desirable. A population has its own specific coadapted gene complex, which may be destroyed by hybridization. The genome of elephants which have evolved in a rain forest may not be adaptive in an arid environment. If a rain forest elephant is introduced into an arid zone, the resulting hybrid offspring may have lower fitness than the original arid-zone elephants (see Templeton (1986) for a discussion on outbreeding depression).

These theoretical considerations of genetic fitness do not preclude the survival of small populations. There is empirical evidence that one small elephant population has increased at a high rate without any apparent genetic abnormality (Hall-Martin 1980). In the Addo National Park, South Africa, there were only 11 surviving elephants in 1931. These increased to 18 by 1953 when the park was fenced, preventing any migration. The population continued to increase, reaching 94 in 1978, at a rate of 7% per year; this rate is very high for an elephant population (Hanks & MacIntosh 1973; Hanks 1979). One reason for the apparent absence of any ill-effects of inbreeding could be the long generation time in elephants. With an increasing generation interval, it would take much longer for inbreeding to accumulate to a certain deleterious level. The mean generation length computed for elephants in the study area is about 18 years. Thus, a constant effective population of 50 elephants would take about 180 years to reach a 10% level of inbreeding. An increase in the population would decrease the rate of inbreeding accumulation.

One exception, however, is no ground for optimism in maintaining very small isolated elephant populations. Ralls *et al.* (1979) have demonstrated that most mammals, including elephants, which are inbred in zoos suffer a higher juvenile mortality than outbred animals. The majority of small wild elephant populations have become extinct over historical periods, even when their habitat was intact. Small populations are always vulnerable to extinction through chance events even if their genetic fitness is not in question.

12.1.2   *Minimum viable area and habitat integrity*

This leads to the related concept of the minimum viable habitat area necessary for the survival of a species. The principles of 'island biogeography' (MacArthur & Wilson 1967) have important implications for the design of terrestrial nature reserves (Wilcox 1980). The rate of extinction of species is usually higher in a smaller 'island' area than in a large area. In an insular habitat the species most vulnerable to extinction are the K-selected ones with low reproductive rates and those at the summit of trophic levels (Terborgh 1976). The question of new colonization of habitat patches, while possible for taxa such as birds, does not usually arise in the case of elephants. It is a matter of either having one large elephant population ranging freely over a large area, or allowing it to break up into a series of smaller populations in fragmented patches. Clearly for the elephant a single large area is desirable.

What would be the minimum viable area for the conservation of the elephant? This is related to the minimum viable population size and to the carrying capacity of the habitat. Assume that the minimum size needed for long-term conservation is 500 breeding individuals; with a prevailing sex ratio of 1 adult male for every 5 adult females, this translates into a total population size of 2200 elephants. If the crude carrying capacity of an area is 0.5 elephant/km^2 and the population is close to this level, a minimum area of 4400 km^2 is needed for its long-term conservation. If the sex ratio is 1:2, the total population size necessary is 1300 elephants. If the density is 0.1 elephant/km^2 as in tropical rain forest habitat, the minimum area required is 13 000 km^2. From a short-term perspective the viable area would be one-tenth of the above figures. The viable area would, therefore, vary with different values of the parameters; the above typical situations give an idea of the scale at which one should think when designing reserves for elephant conservation.

For a species with a large home range and a need for seasonal movement from one habitat type to another, further reduction and fragmentation of the habitat would seriously affect the conservation prospects. Habitat reduction also intensifies the incidence of crop raiding by elephants.

Future land-use planning should aim at preserving habitat contiguity. It would be better to have a single large agricultural project than a series of smaller projects of equal area which would fragment the habitat. Corridors for the free movement of elephants throughout their range should be provided. Where habitats have been severed, a corridor may be provided by raising plantations of suitable trees, including teak and eucalypts. A very long narrow corridor may not be advisable since this would increase the boundary between cultivation and elephant habitat, resulting in crop depredation.

Dams constructed on the periphery of the habitat are less likely to cause fragmentation than those located in the interior. The height of a dam will decide the area of habitat submerged under water. A dam of lower height may provide gently sloping land along the banks of the reservoir which can be used by elephants. When the reservoir is not at its full level, the exposed grassy banks may attract grazers, including elephants (Fig. 12.1). Large reservoirs usually disrupt traditional movement patterns because elephants may not swim across a vast expanse of water. Excessive silt deposition along the banks of a reservoir may prove to be a death trap for elephants coming for water. When canals and pipes associated with reservoirs block the movement of elephants, it should be possible to provide a passage by constructing a bridge across a canal or excavating soil beneath the pipes. A bridge has to be reasonably wide (say, about 5 m) and covered with soil so that elephant herds are willing to use it. The usual experience with narrow bridges over canals in India has been that only adult bulls move across them.

It has been suggested that a disease epidemic is likely to cause the extinction of an entire animal population in a large area. If distributed over numerous smaller habitats the species as a whole would still survive even if one population becomes extinct. This point is valid. But there is no justification for fragmenting a large area if there is an option to retain it intact. The Asian elephant exists in a number of discrete populations spread over a large

Fig. 12.1. The grassy banks of a reservoir may attract herbivores, including elephants, when the water level is low. The picture shows a herd of elephants crossing the reservoir in the famous Periyar Reserve, southern India.

region. It is in that sense buffered against extinction through an epidemic in one region. An effort should be made to keep intact as many large areas as possible along with a number of smaller viable ones. Then, one would not run the risk of 'keeping all the eggs in one basket'.

12.1.3   Maintenance of habitat quality

The habitat has to provide the basic resources of food and water. Undoubtedly, the optimum environment for elephants is one with a diversity of habitat types. This would include moist and dry deciduous forest, scrub thicket, swampy grasslands, riparian forest and patches of evergreen forest. Alluvial floodplains of large rivers are also favoured when associated with a habitat mosaic. This diversity in habitats would enable elephants to optimize their diet depending on seasonal changes in plant phenology. The deciduous forests can be utilized for tall grasses during the wet season. When the grasses become unpalatable during the dry season, elephants need high-protein browse such as legumes found in scrub thicket or grass from swampy grasslands. They need river valleys for water. Riparian forest and evergreen forest also afford them shelter from fires, which are frequent in the dry forests. The highest densities of elephants under natural conditions can be maintained in regions with such a habitat mosaic.

Human use of the habitat need not be always incompatible with the elephant's requirements. Management of the habitat for elephants is tied up with the larger question of sustainable land use. It is not possible to generalize for all areas, but the basic principles of reconciling human needs with those of the elephant are summarized below.

(a) Selective logging in forests is not detrimental. In certain moist vegetation types this may actually create a more favourable niche by increasing the availability of food plants.

(b) Elephants may partly adapt to certain monoculture plantations, such as teak, if the understorey vegetation has sufficient food plants. Weeding should be confined to removing only inedible plants. Monoculture plantations should not be raised on a large scale, because these would decrease the overall habitat diversity.

(c) A check has to be kept on the spread of useless weeds which invade secondary habitats. Such weeds in Indian forests include *Lantana camara* and *Chromolaena* spp. Clear-felled areas are especially susceptible to invasion, but logging or clearing the undergrowth may also allow weeds to establish.

(d) Shifting cultivation on a small scale with a rotational period of over

20 years may be permissible. If cycles are short and the affected area extensive, the practice will desertify the soil and vegetation.

(e) Collection of certain products, such as fruits and grass, may not affect the elephant's resources. Only dead clumps of bamboo should be extracted.

(f) Competition between domestic livestock and elephants may not be serious unless livestock densities are very high. Livestock, however, are competitors with other herbivores such as gaur and deer. The combined herbivore fauna should not consume more than one-quarter of the primary herb layer production (Berwick 1976). There is, however, the danger of livestock transmitting diseases such as anthrax and rinderpest to wild herbivores.

(g) Dry forests susceptible to fire can be managed through annual early cool burning or rotational burning with a period of two to four years. Tall grasses can also be experimentally removed during the dry season as part of fire management. Such habitats have to be monitored for changes in plant species composition due to fire, which may affect food availability. Increase in unpalatable woody plants, such as *Anogeissus latifolia,* or conversion of woodland into grassland would be undesirable. There is urgent need for research on fire ecology before firm recommendations can be made on management.

Should the habitat be purposely manipulated so as to maintain a high carrying capacity for elephants? I personally consider artificial 'management' of natural habitats and animal populations unnecessary except under special circumstances. As argued in Chapter 9, habitat manipulation is likely to increase carrying capacity only in evergreen to moist deciduous vegetation. Tropical evergreen forests possess a unique and diverse assemblage of plants and animals. It is unwise to disturb this community for the purpose of increasing elephant numbers. In the dry deciduous and xerophytic vegetation, any conversion to secondary forms will not benefit the elephant.

Where elephants range over a sufficiently large area, it is not necessary to manipulate the habitat even if they exist at only a low density. Animal populations must be allowed to regulate their numbers naturally in relation to environmental factors. An artificial increase in the elephant population will result in increased crop raiding, damage to woody vegetation and a crash in numbers during droughts. Some artificial management may be justified if elephants are confined to a small area, in order to maintain a viable population size.

210   Conservation and management

12.1.4   *Reduction of poaching*

The illegal ivory trade has to be controlled in certain regions such as southern India where poaching is rampant. Anti-poaching squads of the Forest Department employing local people have been partly successful in Karnataka state. An intelligence network has proved to be more effective in identifying and apprehending the key persons in the poaching trade.

A realistic policy on the ivory trade has to be formulated. The main thrust of this policy should be to decrease the economic value of ivory, and hence make poaching less attractive. The high rate of customs duty on imported African ivory has been responsible for the high value of local ivory in India. If legal African ivory is made available at cheaper rates, the carvers may reduce patronage of poached ivory. At present the ban on sale of tusks recovered by the Forest Department from naturally dead elephants has further reduced the supply to carvers. This policy is unrealistic because there is no method as yet to differentiate between Asian and African elephant ivories. Substitutes for ivory should be encouraged. One synthetic product named ivorine seems to have partial acceptance. Technology for such products should be acquired by countries like India where use of ivory products (such as bangles by brides) is rooted in traditional culture. It would, however, be naive to expect synthetic ivory to totally replace real ivory, just as imitation gold has not replaced real gold.

12.1.5   *Captive breeding programmes*

Domestic stocks of elephants are bound to decline if not replenished through captures from the wild. There is scope for slowing the decline through captive breeding programmes. Although no successful captive breeding model is available it is surely possible to develop one through investigations into the reproductive biology of captive elephants. Both social and physiological factors may be responsible for the poor birth rate among domestic elephants. A large proportion of captive elephants are scattered in small groups or as solitaries. These may never have an opportunity for reproduction. On the other hand, elephants in southern India held by the Forest Department under semi-natural conditions have been fairly successful breeders. These elephants are kept in larger groups (usually more than 10 animals) and let out for feeding at night into the forest, where the cows may be mated by wild bulls. Captive bulls also sire many of the calves born. Between 1950 and 1983 about 74 calves were born to 37 captive adult cow elephants in Tamilnadu state. Although some adult cows have calved regularly, others have never reproduced even though maintained under similar conditions. Causes of this apparent sterility have to be

12.2 Protection of human interests

This section discusses various means of protecting human life, crops and property from elephants.

12.2.1 *Elephant-proof barriers*
(a) Trenches

A trench has to be at least 2 m deep, 2 m across at the top and 1.5 m across at the base in order to effectively prevent elephants from crossing over. Trenches have a high rate of failure especially in wet conditions when the soil is loose. Elephants dig the soil with their forefeet, partly fill up the trench and then get across. All the trenches inspected in southern India had failed to keep out adult bull elephants. Trenching has been given up by the Federal Land Development Authority in Malaysia (Blair *et al.* 1979).

The cost of manually digging a trench of the above dimensions is U.S.$2500 per km (at 1982 prices) in southern India and approximately the same using mechanical excavators, in Malaysia. Improved trenches reinforced with stones, tar or latex would cost up to five times the amount.

(b) High-voltage electric fence

Electrified fences of a non-fatal type have been used to keep a variety of mammals, including predators, away from waterfowl nests (Lokemoen *et al.* 1982) and elephants from plantations in Malaysia (Blair *et al.* 1979). Detailed fence designs are provided elsewhere (Blair *et al.* 1979; Piesse 1982; INPUT 1983). Only a brief account of the fence design which can be used against elephants is given here.

The fence consists of one or more strands of high-tensile (100 tonnes tensile) galvanized steel wires strung at appropriate heights above the ground on hardwood posts (Fig. 12.2). For elephants, at least two strands at heights of about 1 m and 2 m above ground should be used. The posts are protected by vertical wires. Porcelain insulators may be provided at places where the wires come into contact with the posts. The posts may be spaced up to 20 m apart for a wire tension of 180 kg. The ground below the wire should be free of vegetation to prevent leakage of current. The heart of the fence is the 'energizer' (transformer), which typically gives an electric pulse of 5000 V and 1/3000 second duration every second. Owing to the high voltage (which is in practice 3000 V along the lines), an animal coming into contact with the

212 Conservation and management

Fig. 12.2. A high-voltage electric fence in the Nilgiris, southern India. The picture below shows the energizers which are used to regulate the current.

wire receives a severe shock but the short duration of the pulse prevents any injury. The energizer may be powered by a 12 V car battery, from the 110 V/230 V mains or even by solar cells. Each energiser can effectively power about 20 km of fencing.

The costs of material and labour for the fence are between $400 and $800 per km depending on the sophistication of design. A standard energizer costs $250 to $500. Maintenance of the fence would annually cost 10–20% of the capital.

Numerous experiments with elephants in Africa (R. L. Piesse, unpublished results) and Malaysia have shown that electric fencing is generally effective. During a period of 32 days in 1982, a total of 259 elephants made 184 contacts with an electric fence in Namibia–Etosha, but not a single elephant got through. In Malaysia, a few thousand kilometres of fencing have been erected around oil palm and rubber plantations, for which a success rate of 80% has been reported. Farmers in the Nilgiris of southern India have also begun to use electric fencing. Most instances of failure were due to faulty design and improper maintenance. Elephants may also learn the weaknesses of the fence. One bull elephant consistently broke through a fence in Kemasul, Malaysia, by rearing up on its hind legs, placing its front foot on the upper wire (which sometimes snapped) and crossing over. The sole of the foot is a bad conductor. Another trick that a bull may try is to use its tusks (a non-conductor) to break the wire or prise an insulator loose (modified designs do not use insulators). It must be emphasized that the electric fence is not a physical barrier but merely a psychological bluff.

(c) Other barriers

Many other barriers have been suggested, including a 'sausage barrier' of boulders held together by wire netting, a rubble wall along a trench, a fence of iron pipes and a spike barrier using galvanized nails. The prohibitive cost of these barriers is a constraint on their use over a large area.

If dams are constructed on the periphery of elephant habitat, it may be possible to route irrigation canals and pipes along the boundary of forest and cultivation. These would allow elephants complete access to their natural habitat but prevent them from entering cultivation. The contour canal from Sirkarapathy power house to Aliyar reservoir in the Anamalai hills of southern India is an example of such a barrier. The Transbasin canal of the Mahaweli project in Sri Lanka performs a similar function.

(d) Cost-effectiveness of barriers

The cost of a trench is about three times that of a good electric fence. For instance, a 10 km trench around the study village of Hasanur would cost

$25 000, compared with about $7500 for a 2-strand electric fence. Maintenance of the fence would cost about $1500 annually, including materials, labour and wages for a part-time electrician. The damage to crops amounted to $5700 at Hasanur in 1981. The capital costs of fencing could be recovered within two years. The value of crops saved annually would be about three times the recurring costs of maintenance, assuming 80% reduction in damage. In larger villages suffering lesser damage the cost-effectiveness would be lower. Trenches would be much less effective considering their capital costs, maintenance and rate of failure.

Electric fences operating in Malaysia cost $1200 per km for construction and $1600 for maintenance over a period of five years. Assuming that 1 km of fencing of 80% efficiency protects 200 ha of oil palm plantation and that 10% of the trees with an average age of 21 months are potentially subject to damage by elephants annually (Blair *et al.* 1979), the total export value of crops saved would be $208 000. That is, for every dollar invested in fencing about 74 dollars are saved over five years.

The cost-effectiveness of the electric fence is certainly much higher for cash crops such as oil palm and rubber compared to millet crops. However, if many other factors are considered in the calculation the difference is not that large. The wage potential of man hours of labour spent in guarding fields at night would be considerable in southern India. Farmers do not cultivate a significant proportion of land close to the forest boundary because of their fear of depredation by elephants. Any scheme to prevent crop damage should be combined with an integrated programme of increasing irrigation, productivity and area of land under cultivation, planting fuel wood and fodder trees, stall-feeding of cattle and other eco-development measures. One argument in favour of trenching is that the money spent goes entirely to poor labourers. But surely it is possible to engage them in more productive work than merely digging soil, such as planting trees. One problem with the electric fence is that the materials used may be stolen. An electric fencing project which protects numerous land owners should be taken up as a co-operative scheme. The ultimate responsiblity of maintaining the fence should rest with the people whose land it protects.

12.2.2 *Discouraging and chasing elephants*
(*a*) Buffer zones
Seidensticker (1984) recommended the creation of 'wide buffer zones', which lack any cover, between cultivation and natural habitat. The idea is to discourage elephants from using these areas, and hence avoid elephant contact with cultivation. This method cannot be applied to most regions.

Such buffer zones will result in further loss of natural habitat. Further, adult bull elephants would certainly use the cover of darkness to cross these buffer zones and enter cultivation. Habitual crop-raiding elephants cannot be kept away by such simple bluffs.

(b) Chemicals
Spraying of chemicals which may repel elephants has been suggested. No chemical has been proved effective so far. Chemicals may not persist in the environment under wet conditions.

(c) Sound
The use of high-frequency sound 'beepers' along with an electric fence may have potential in repelling elephants (Piesse 1982). Further research is needed in identifying a noise frequency intolerable to elephants.

(d) Patrols
Patrols along the forest boundary at night by people armed with guns (to shoot in the air), flashlights and noise-making devices may be partly effective. In addition to being a tiresome job, this does not work against experienced raiding elephants.

(e) Anchored mela-shikar
Trained domestic elephants were successfully used in chasing away a herd of about 60 elephants which was raiding crops in West Bengal state during November 1980 (D. K. Lahiri Choudhury, personal communication). This type of chase without capture has been termed anchored *mela-shikar*.

12.2.3   *Culling, capture and translocation of elephants*
Notorious crop-raiding elephants or small isolated populations which are in regular conflict with people will have to be eliminated, captured or translocated. The Asian tradition of domesticating elephants makes it possible for herds to be captured rather than killed. Herds may also be translocated to a suitable habitat. There may be no alternative to eliminating adult male elephants which have turned 'rogues' and are dangerous to domesticate.

(a) Selective removal of male elephants
The removal of adult bulls which are habitual crop raiders or killers will reduce depredation and save human lives to a far greater degree than elimination of elephants from family herds (Chapters 7 and 8). It is also

relatively easy to identify a notorious bull. Rogue bulls will usually have to be shot. Crop raiders may be chemically immobilized and captured for domestication. In regions where the adult sex ratio is not very disparate, the elimination of some bulls will not have any adverse effect on population demography. A positive population growth rate can be maintained even with an adult ratio of 1 male for every 5 adult females (Chapter 11). Most elephant populations in Asia except those in southern India (and, perhaps, Burma) are likely to have a more equal sex ratio. Populations in Sri Lanka and northeastern India with a large proportion of tuskless males, immune to poaching, may have 'surplus' males which can be selectively culled if necessary. Elimination of male elephants may not be justified if they are already under severe poaching pressure, as in southern India.

(b) Use of drugs in capture

The technique of chemical immobilization perfected on the African elephant has helped numerous research studies (e.g. Elder & Rodgers 1974; Douglas-Hamilton & Douglas-Hamilton 1975; Leuthold 1977; Rodgers & Elder 1977). It is increasingly becoming a management tool in Asian countries for dealing with problematic elephants. Solitary bulls or small elephant groups can be conveniently captured by means of drugs. The use of drugs requires an adequate understanding of drug-administering devices, the pharmacology of drugs and the physiology of the animal. They should therefore be handled only by trained personnel. There are several standard references dealing with chemical immobilization (e.g. Harthoorn 1976). A brief account of its application to elephants is given here.

A Chap-Chur syringe dart is the most common device used to load the drug. On making contact with the animal a 0.22 calibre blank explosive fires, forcing the plunger to the front and injecting the drug into the animal. Of several types of propellant guns, a crossbow was preferred by Elder & Rodgers (1974) for its accuracy and silent delivery of the dart. The drug most commonly used to immobilize African elephants is etorphine hydrochloride (M-99), an extremely potent narcotic which acts as a depressant upon certain central nervous system functions. The effective dose of intramuscular etorphine which induces immobility in elephants within 10 minutes is about 180 µg/kg of body weight (Elder & Rodgers 1974). The body weight can be predicted from measurement of shoulder height or circumference of front foot (Appendix II). Once an elephant is down it is dangerous to keep it in sternal recumbency; it should be rolled into lateral recumbency. About 16–25 mg of the antidote, cyprenorphine hydrochloride (M-285), revives elephants in three minutes on average.

Commercially available Immobilon is a combination of one part of etorphine with four parts of acepromazine (acetylpromazine maleate). Immobilon can be dissolved in saline such that 1 ml of fluid contains 12.5 mg of the drug (2.45 mg etorphine and 10 mg acepromazine). Olivier (1978b) used 0.4 ml of Immobilon per foot of shoulder height of the elephant. Eight bull elephants in the isolated Kattepura Reserve of Karnataka state were captured during March 1987 by using Immobilon (M. K. Appayya, personal communication). The antidote used was diprinorphine (Revivon) in quantities matching the dosage of Immobilon. Before loading the elephants onto trucks, they were put under sedation by injecting 100–150 mg each of xylazine (Rompun) and ketamine.

The use of etorphine is strictly regulated in many countries because it is highly dangerous to people. In the state of Kerala, numerous domestic elephants which ran amuck have been brought under control by tranquillizing them with a combination of xylazine and acepromazine (J. V. Cheeran, personal communication). Typically, about 10 mg of xylazine was combined with 2.5–5 mg of acepromazine for every 100 kg body weight of the elephant. Owing to the long induction period, these chemicals may be of limited use for capturing wild elephants.

(c) Traditional methods of capture

In spite of the development of chemical methods of capture, there is still an important role for traditional methods in elephant management. Chemically immobilized elephants are best handled after revival using trained domestic elephants or *koonkies*. The only way of capturing a large family herd is through the *kheddah* method of driving the elephants into a stockade. The pit method may have some limited use in trapping crop raiding elephants which enter cultivation along specific paths from the forest. *Mela-shikar* or noosing an elephant from a captive elephant's back may be used in capturing young bulls.

(d) Translocation through elephant drives

Isolated elephant herds can often be driven to a larger habitat. One of the first successful elephant drives was carried out in Sri Lanka during 1978–79. About 160 elephants in system H of the Mahaweli Scheme, which was to be submerged, were driven to the Wilpattu National Park, a distance of over 50 km, without any casualties to either people or elephants. This involved bringing the elephants through cultivation, across roads and channels and keeping them 'boxed-in' at selected sites at night. Another successful translocation was that of about 70 elephants from the Gunung Madu Sugar Cane

Plantations to the Way Kambas Game Reserve in Sumatra during 1984–85 (Santiapillai & Suprahman 1985). Traditional expertise in conducting *kheddahs* can be utilized for such translocations.

12.2.4 Agricultural planning

If all farmers within an area cultivate a seasonal crop at the same time, the overall damage can be reduced. Asynchrony increases the total period of cultivation and, hence, the period of raiding by elephants. Alternative cultivation patterns may succeed if these are financially remunerative. Certain oil seed crops such as niger (*Guizotia abyssinica*) and gingelly (*Sesamum indicum*), which are not consumed by elephants, are potential replacements for millets in southern India. It may be feasible to translocate settlements from elephant habitat declared as national parks. Isolated patches of forest outside the park, which cannot be utilized by elephants, would be suitable sites for relocating settlements. Since translocation is an expensive exercise, it may be possible to recover a part of the costs by raising plantations of fast-growing trees on the released land before allowing it to revert to wilderness.

12.2.5 Social security schemes

Farmers in some regions are eligible to receive compensation from the government for damage to their crops by elephants. Crop insurance schemes can also be tried out. Similarly, loss of human life should be adequately compensated, especially if elephants kill people within human settlement. Under an insurance scheme suitable for the poor in India, an annual premium of $1.00 provides a cover of $1200 against accidental death. Such social security schemes are necessary to make people take a favourable attitude towards conservation, until more permanent solutions to the problem are implemented.

APPENDIX I
Estimation of seasonal elephant densities by ground transects

Each zone was taken as the basic unit for sampling elephant density. Every trip into the field, by vehicle or on foot, was considered as a transect sample. Road transects are often likely to be biased. This was largely minimized because the study area was divided into a number of habitat types for the purpose of sampling. Within any zone there was no obvious bias in the nature of roads; for instance, roads in most hilly zones covered valleys, slopes and hill tops. The distance travelled, time spent and the number of elephants seen in each zone were recorded. For each zone the visibility was measured along the commonly used roads at intervals of 0.2–0.5 km so as to sample at least 30 locations. The distances to which an elephant could be spotted on both sides of the road were noted at each location. The average visibility for each zone was estimated at different seasons.

Ideally, if an observer were to transect an area as quickly as possible (either by aircraft or on the ground by vehicle), the elephants recorded would be only those present at a given moment within the area visible. If the observer waited for some time, then more elephants could be expected to come into view. This 'flow' would be proportional to the time spent and the density of elephants. It has to be assumed that the average rate of movement of elephants is roughly the same in different zones since the observer has no control over this factor. Thus, the density of elephants calculated from transects involving a waiting period is likely to be greater than the real density. One solution to this problem would be to consider only those elephants seen while travelling and ignore those seen when the observer is stationary. But this would reduce sampling intensity in those zones where the distance travelled was short but in which the time spent was considerable. The problem lies in suitably incorporating the area scanned and the time spent into the model. Since area and time are separate dimensions it is not possible easily to relate one in terms of the other.

Zones No.	Area (km²)	1981					1982						1983
		M-A	M-J	J-A	S-O	N-D	J-F	M-A	M-J	J-A	S-O	N-D	J-F
1	55	A	C	C	C	D	B	A	D	C	C	D	C
2	32	A	A	B	B	C	B	A	A	A	A	A	A
3	125	1.0	1.1	E	C	B	0.6	0.9	0.8	1.2	C	0.2	0.6
4	78	A	A	A	A	A	A	A	A	A	A	A	A
6	37	E	E	D	E	D	D	E	E	E	E	D	D
7	60	C	1.4	1.9	2.5	3.0	C	0.4	0.7*	E	E	E	B
9	21	A	<0.1	A	B	0.2	A	A	A	<0.1	B	C	A
10	53	B	B	B	D	D	B	B	B	D	D	D	D
11A	62	B	D	D	B	C	B	B	D	C	C	C	A
11B	96	1.3*	0.4*	0.2	0.1	0.7*	0.5*	0.2*	0.2	0.1*	B	1.6*	1.6*
12	27	4.3*	0.8*	0.7*	0.8*	0.5*	5.0*	3.7*	0.6	0.8*	D	0.3*	1.3*
13	28	C	A	A	C	B	C	B	A	A	C	B	A
16	12	1.3*	1.6*	C	0.8	1.7*	3.6*	0.4*	0.5	1.2*	E	2.0*	2.4*
17	136	<0.1*	<0.1*	<0.1*	A	<0.1	<0.1	<0.1	<0.1	<0.1	A	<0.1	0.2*
18	44	B	<0.1	A	A	E	F	C	A	A	B	0.8	1.9
19	62	D	B	B	C	2.5	1.4	0.8	B	B	D	2.6	3.0*
Mean density		0.61	0.50	0.52	0.45	0.74	0.68	0.46	0.39	0.49	0.44	0.70	0.75
Number of elephants		564	463	483	420	691	628	424	364	452	412	654	699

The actual densities in elephant/km² are given for instances of high sampling (indicated by *) and medium sampling efforts.

The density classes for low sampling effort are: A = <0.1, B = 0.1–0.25, C = 0.25–0.5, D = 0.5–1.0, E = 1.0–2.0, F = 2.0–4.0 elephants/km².

Evidence that the density estimates were reasonably accurate came from a total census count carried out by the Forest Department on 30 April 1983. This census gave a tally of 691 elephants for the 928 km² sampled study area, a density of 0.74 elephant/km². This compares well with the density of 0.70 elephant/km² for November–December 1982 and 0.75 elephant/km² for January–February 1983, estimated from the transect data. Because sampling intensity towards the end of the study was insufficient during March–April 1983 in many zones, it is not possible to make a direct comparison with the Forest Department's census data.

Taken from Sukumar (1989*a*).

Appendix I

For the present study a relatively simple approach was taken. For one hilly habitat (Zone 11B) and one valley habitat (Zone 12) the records of elephant sightings during the two-month periods of highest sampling intensity were separated into two categories: those seen while the observer was in motion (distance travelled and area scanned is available) and those sighted when stationary (total time spent is available). The two-month periods were March–April 1981 for Zone 11B (258 km travelled, 32 hours spent) and March–April 1982 for Zone 12 (426 km travelled, 54 hours spent). From these it was calculated that the number of elephants seen for every 1 km² area scanned was the same as the number seen for about 3 hours (3.0 in Zone 11B and 2.8 in Zone 12) spent in waiting. Such a relation was determined for only these two representative zones, as the very high sampling effort and relatively high elephant density ensured the best possible estimates for the relationship. It has been assumed that this relation also holds good for other zones. The formula used in calculating the seasonal density of elephants in different zones was

$$D = \frac{N}{A + \frac{1}{3}T},$$

where D = density of elephants in km²
N = total number of elephants seen
A = area scanned in km²
(area = distance travelled × mean visibility)
T = time spent in hours.

The estimates of elephant density are grouped into two-month classes for different zones. The results have been expressed at three levels of sampling intensity.

(a) *High sampling effort*: The area scanned exceeded 30% of the zone area, or over 100 km travelled or over 30 hours spent in a particular zone during a two-month period. The actual densities (indicated by asterisks) are given.

(b) *Medium sampling effort*: The area scanned exceeded 15% of the zone area, or over 50 km travelled or over 15 hours spent. The actual densities are given.

(c) *Low sampling effort*: Sampling intensities were below the medium level. Only the density ranges are given.

Subjective estimates, based on careful assessment of indirect signs of elephant presence, have been made in some zones where sampling intensity was low or which were not sampled during a particular season. In many cases this would not introduce any serious error into the overall data because elephant densities in these zones (such as Zones 2, 4 and 13) were clearly very low. For instance, in Zone 13 few elephants would have utilized the steep slopes with rocky cliffs, which in any case were scanned regularly from Zone 12.

The density estimates refer only to day-time occupancy. In certain zones the densities could change during the night. Few elephants occupied Zone 9, which is a series of *Eucalyptus* plantations, during the day, although this zone was used more frequently at night as a passage to cultivation or to move between Zones 8 and 10.

APPENDIX II
Growth relationships and field methods of ageing elephants

Age criteria based on dentition (Laws 1966; Lang 1980) have been used to obtain age structures of culled African elephant populations (see, for example, Hanks 1972*a*; Laws *et al.* 1975; Smuts 1977). Less accurate field methods of ageing, based on the relationship between age and body size attributes such as height or length (Laws 1966; Hanks 1972*b*; Laws *et al.* 1975), have also proved useful in constructing age structures (see, for example, Douglas-Hamilton 1972; Croze 1972; Jachmann 1980). Methods of estimating body size by photography are improvements over purely subjective assessments of size in the field (cf. McKay 1973; Kurt 1974) which cannot be verified.

(a) Measuring height by photography
The principle of measuring parallax differences by stereo photography has been used by Douglas-Hamilton (1972) and Hall-Martin & Ruther (1979) to estimate the height of elephants. Conventional photography can also be used if the position of the elephant at the time of taking the picture can be accurately determined. A calibrated pole can be held by an assistant at the same spot as the elephant (after it has moved away!) and photographed from the original position (Jachmann 1980). The elephant's height can be determined by reference to this scale.

In this study, a modified pole method was used in estimating height. Elephants were photographed when they crossed a road in a single file or came along a favourite path to a water hole. The distance from my position to that of the elephant was measured by a tape. All the negatives were uniformly magnified 10.7 times for better resolution in measuring height on the prints. A pole was independently calibrated at distances from 25 to 100 m at intervals of 5 m. A lens of 200 mm focal length was used throughout. Heights were calculated from the standard lens equations and by reference to the calibrated pole. Details of the calculations are available elsewhere (Sukumar 1985 and unpublished manuscript). A test of accuracy carried out on 10 captive elephants of known height gave a mean deviation of 3.1% (range 1.1–5.3%). There was less than 4% difference in heights calculated for some identified adult wild elephants photographed at different times.

In some instances, when members of a herd were standing close to each other and the distance of photography could not be measured, the heights of sub-adult elephants were estimated by relating these to an adult female elephant. The height of a large cow was taken to be the asymptotic height (240 cm) and others in the group were proportionately graded from measurements on the print.

Age and growth parameters in Asian elephants

Age (years)	Height (cm) M	Height (cm) F	Weight (kg) M	Weight (kg) F	Tusks in males CTLL (cm)	Tusks in males Weight (kg)
0	90	89	120	120	—	—
1	121	119	330	310	—	—
2	139	135	520	470	—	—
3	155	149	705	610	7.6	0.1
4	169	161	920	710	9.8	0.2
5	180	170	1130	810	11.9	0.4
6	190	177	1340	930	13.8	0.7
7	198	183	1540	1055	15.7	1.1
8	205	188	1730	1180	17.4	1.5
9	212	193	1900	1300	19.0	2.1
10	217	197	2065	1415	20.5	2.6
11	222	200	2200	1525	21.9	3.2
12	225	203	2320	1635	23.3	3.9
13	228	206	2400	1735	24.5	4.6
14	231	209	2500	1830	25.7	5.3
15	235	213	2645	1925	26.8	6.1
20	250	228	2970	2300	31.3	10.0
25	262	234	3400	2560	34.6	13.6
30	268	238	3650	2740	37.0	16.8
40	272	240	3800	2930	40.0	21.4
∞	274	240	3900	3000	43.4	27.4

The height is twice the circumference of the front foot (CFF) for all ages.
Weights in juvenile animals are based on small sample sizes.

(b) Growth relationships

A large data base on captive Asian elephants (kept under semi-natural conditions by the Forest Departments in southern India) was used to fit von Bertalanffy growth equations using a computer (von Bertalanffy 1938; Laws 1966; Hanks 1972b; Laws *et al.* 1975). The growth parameters analyzed included height, circumference of front foot (CCF), body weight and circumference of tusks at lip line (CTLL) versus age. Relationships between height and weight, CTLL and tusk weight, CFF and height, etc. were also determined (Sukumar 1985; Sukumar, Joshi & Krishnamurthy 1988). Suitable corrections were made to the data because there was evidence for a depressed post-pubertal growth in captive elephants compared with wild elephants. These growth parameters were used to assign ages to living and dead elephants, estimate biomass of the study populations, etc. A reference table on growth parameters is provided on p. 225.

(c) Ageing adult elephants

It is possible to age male elephants up to 20 years and female elephants up to 15 years from their heights. As the height approaches the asymptotic value it cannot be used for ageing. Adult elephants were therefore placed in age classes by comparing certain morphological features with those of known-aged captive elephants.

Albl (1971) has shown that the buccal depression and the temporal dent are more indicative of age than of physical condition in African elephants. Although some deepening occurs during the dry season because of reduced fat deposits, the main trend is increased depth with increasing age.

The degree of turnover of the ear flaps is also indicative of age. In Asian elephants the turnover is forwards. The turnover usually takes place between 20 and 40 years of age.

A number of known-aged captive elephants were examined and photographed. The buccal depression, temporal dent and ear turnover in wild elephants were compared with those of captive elephants for assigning age classes.

APPENDIX III
Nutritive value of food plants

A – Crude protein

Plant species	Plant parts	Month and season	Crude protein % dry matter
Wild browse plants			
Kydia calycina	leaves	Feb., dry	8.8
	leaves	Dec., II wet	15.6
Grewia tiliaefolia	leaves	Feb., dry	10.4
	leaves	Aug., I wet	14.5
	leaves	Dec., II wet	13.0
	bark	Aug., I wet	3.6
Ziziphus xylopyrus	leaves	Feb., dry	7.6
Acacia pennata	leaves	Feb., dry	17.6
Acacia torta	leaves	Mar., dry	11.3
	leaves	Aug., I wet	14.9
Acacia suma	leaves	Jan., dry	17.7
	leaves	Aug., I wet	25.5
Acacia leucophloea	leaves	Mar., dry	12.0
Albizia amara	leaves	Feb., dry	18.0
Phoenix humilis	leaves	Feb., dry	6.0
Bambusa arundinacea	leaves	May, I wet	14.5
Dendrocalamus strictus	leaves	Dec., II wet	14.4
Wild grasses			
Themeda cymbaria	leaves	Feb., dry	3.2
	bases	Feb., dry	1.6
	leaves	May, I wet	9.9
	leaves	June, I wet	9.5
	leaves	Aug., I wet	9.3
	bases	Aug., I wet	2.2

continued

A – Crude protein, continued

Plant species	Plant parts	Month and season	Crude protein % dry matter
Themeda cymbaria	leaves	Nov., II wet	8.0
cont.	bases	Nov., II wet	3.8
Themeda triandra	leaves	Feb., dry	3.8
Cymbopogon flexuosus	leaves	Feb., dry	4.8
	bases	Feb., dry	2.5
	leaves	June, I wet	8.4
	bases	June, I wet	3.2
	leaves	Nov., II wet	6.8
	bases	Nov., II wet	2.0
Cultivated grasses			
Sorghum vulgare	vegetative stage	June, I wet	13.0
	inflorescence stage	July, I wet	11.6
Zea mays	vegetative stage	June, I wet	7.9
	entire cob	July, I wet	12.0
Eleusine coracana	vegetative stage	Sept., II wet	11.2
	inflorescence stage	Oct., II wet	8.3
	grain stage	Nov., II wet	5.3
Oryza sativa	inflorescence & grain stage	Nov., II wet	10.4
	basal portion of stem	Nov., II wet	9.8

B – mineral content

Plant species	Plant part	Month and season	Total ash	Calcium	Sodium	Magnesium	Iron
Wild browse plants							
Kydia calycina	bark	Feb., dry	80.0	24.6	0.15	1.05	0.04
	bark	July, I wet	176.6	57.2	0.16	2.12	0.20
	bark	Dec., II wet	75.4	26.5	0.13	1.19	0.19
Grewia tiliaefolia	bark	Feb., dry	74.4	22.6	0.20	0.99	0.06
	bark	Dec., II wet	98.7	27.6	0.17	1.28	0.07
Ziziphus xylopyrus	leaves	Feb., dry	51.1	17.7	0.11	0.83	0.09
	bark	Feb., dry	102.4	37.4	0.12	1.37	0.20
Acacia pennata	bark	Feb., dry	68.2	21.8	0.17	0.83	0.07
Acacia torta	bark	Mar., dry	66.4	19.0	0.15	0.88	0.02
Acacia suma	bark	Jan., dry	58.3	17.9	0.18	0.74	0.04
	bark	Aug., I wet	63.2	?	0.18	0.76	0.28
Acacia leucophloea	bark	Mar., dry	79.5	23.5	0.14	0.99	0.03
Bambusa arundinacea	leaves	May, I wet	96.0	2.9	0.27	1.14	0.48
Dendrocalamus strictus	leaves	Dec., II wet	64.5	2.5	0.13	0.79	0.11
Wild grasses							
Themeda cymbaria	leaves	Nov., II wet	73.7	4.6	0.13	0.80	0.14
	bases	Nov., II wet	52.3	2.3	0.16	0.59	0.08
Cymbopogon flexuosus	leaves	Nov., II wet	81.2	1.9	0.12	0.77	0.34
	bases	Nov., II wet	75.1	0.8	0.28	0.57	0.43
Cultivated grasses							
Sorghum vulgare	inflorescence stage	July, I wet	42.5	0.8	0.10	0.53	0.06
Zea mays	entire cob	July, I wet	60.2	0.9	0.25	0.75	0.10
Eleusine coracana	vegetative stage	Sept., II wet	105.1	7.6	0.57	1.06	0.18
	inflorescence stage	Oct., II wet	73.7	10.8	0.94	1.00	0.25
Oryza sativa	grain stage	Nov., II wet	47.9	7.4	0.24	0.66	0.09
	inflorescence & grain stage	Nov., II wet	160.1	2.6	0.36	1.26	0.21

All values are in mg/g dry matter.
Taken from Sukumar (1989b).

REFERENCES

Albl, P. (1971). Studies on assessment of physical condition in African elephants. *Biological Conservation* **3**, 134–40.

Alexis, L. (1984). Sri Lanka's Mahaweli Scheme – The Damnation of Paradise. *The Ecologist* **14**, 206–15.

Allaway, J. D. (1979). Elephants and their interactions with people in the Tana river region of Kenya. Unpublished Ph.D. thesis, Cornell University, Ithaca.

Allendorf, F. W. & Leary, R. F. (1986). Heterozygosity and fitness in natural populations of animals. In *Conservation Biology,* ed. M. E. Soulé, pp. 57–76. Sunderland, Massachusetts: Sinauer Associates.

Altmann, J. (1974). Observational study of behaviour: Sampling methods. *Behaviour* **49**, 227–67.

Ananthasubramaniam, C. R. (1980). A note on the nutritional requirements of the Asian elephant (*Elephas maximus indicus*). *Elephant* (Supplement) **1**, 72–3.

Andau, P. & Payne, J. (1985). Elephants in Sabah. *WWF Monthly Report*, December 1985, pp. 301–5. Gland, Switzerland: World Wildlife Fund.

Anderson, G. D. & Walker, B. H. (1974). Vegetation composition and elephant damage in the Sengwa Wildlife Research Area, Rhodesia. *Journal of the South African Wildlife Management Association* **4**, 1–14.

Balfour, E. (1885). *The Cyclopaedia of India*, vol. 1. London: Bernard Quaritch.

Barnes, R. F. W. (1980). The decline of the baobab tree in Ruaha National Park, Tanzania. *African Journal of Ecology* **18**, 243–52.

Barnes, R. F. W. (1982). Elephant feeding behaviour in Ruaha National Park, Tanzania. *African Journal of Ecology* **20**, 123–36.

Bax, P. N. & Sheldrick, D. L. W. (1963). Some preliminary observations on the food of elephants in the Tsavo Royal National Park (East) of Kenya. *East African Wildlife Journal* **1**, 40–53.

Bell, R. H. V. (1971). A grazing ecosystem in the Serengeti. *Scientific American* **224**, 86–93.

Bell, R. H. V. (1981). Notes on elephant–woodland interactions. In *The Status and Conservation of Africa's Elephants and Rhinos,* ed. D. H. M. Cumming & P. Jackson, pp. 98–103. Gland, Switzerland: International Union for Conservation of Nature and Natural Resources.

Belovsky, G. E. (1981). Food plant selection by a generalist herbivore: The moose. *Ecology* **62**, 1020–30.

Belovsky, G. E. (1984). Herbivore optimal foraging: A comparative test of three models. *American Naturalist* **124**, 97–115.

Benedict, F. G. (1936). *The Physiology of the Elephant*. Carnegie Institute of Washington Publication No. 474.

Berwick, S. (1976). The Gir forest: An endangered ecosystem. *American Scientist* **64**, 28–40.

Blair, J. A. S. (1980). Management of the 'Agriculture–Elephant interface' in peninsular Malaysia. Paper presented at the second meeting of the IUCN/SSC Asian Elephant Specialist Group, Colombo.

Blair, J. A. S., Boon, G. G. & Noor, N. M. (1979). Conservation or cultivation: The confrontation between the Asian elephant and land development in peninsular Malaysia. *Land Development Digest* **2**, 27–59.

Blouch, R. A. & Haryanto (1984). Elephants in southern Sumatra. Report No. 3, World Wildlife Fund Project 3133, Bogor.

Blouch, R. A. & Simbolon, K. (1985). Elephants in northern Sumatra. Report No. 9, World Wildlife Fund Project 3133, Bogor.

Blower, J. (1985). Elephants in Burma. *WWF Monthly Report,* December 1985, pp. 279–85. Gland, Switzerland: World Wildlife Fund.

Botkin, D. B., Mellilo, J. M. & Wu, L. S.-Y. (1981). How ecosystem processes are linked to large mammal population dynamics. In *Dynamics of Large Mammal Populations,* ed. C. W. Fowler & T. D. Smith, pp. 373–87. New York: John Wiley & Sons.

Buechner, H. K. & Dawkins, H. C. (1961). Vegetation changes induced by elephants and fire in Murchison Falls National Park, Uganda. *Ecology* **42**, 752–66.

Burne, E. C. (1942). A record of gestation periods and growth of trained Indian elephant calves in the southern Shan States, Burma. *Proceedings of the Zoological Society of London* (Series B) **112**, 27.

Buss, I. O. (1961). Some observations of food habits and behaviour of the African elephant. *Journal of Wildlife Management* **25**, 131–48.

Buss, I. O. (1977). Management of big game with particular reference to elephants. *Malayan Nature Journal* **31**, 59–71.

Caras, R. A. (1975). *Dangerous to Man*. London: Barrie & Jenkins.

Carrington, R. (1958). *Elephants*. London: Chatto & Windus.

Caughley, G. (1974). Interpretation of age ratios. *Journal of Wildlife Management* **38**, 557–62.

Caughley, G. (1976). The elephant problem – An alternative hypothesis. *East African Wildlife Journal* **14**, 265–84.

Caughley, G. (1977). *Analysis of Vertebrate Populations*. London: John Wiley & Sons.

Caughley, G. (1981). What we do not know about the dynamics of large mammals. In *Dynamics of Large Mammal Populations,* ed. C. W. Fowler & T. D. Smith, pp. 361–72. New York: John Wiley & Sons.

Caughley, G. & Lawton, J. H. (1981). Plant–herbivore systems. In *Theoretical Ecology: Principles and Applications*, ed. R. M. May, pp. 132–66. Oxford: Blackwell Scientific.

Clemens, E. T. & Maloiy, M. O. (1982). The digestive physiology of three East African herbivores: The elephant, rhinoceros and hippopotamus. *Journal of Zoology, London* **198**, 141–56.

Conn, E. E. (1979). Cyanide and cyanogenic glycosides. In *Herbivores: their Interactions with Secondary Plant Metabolites,* ed. G. A. Rosenthal & D. H. Janzen, pp. 387–412. New York: Academic Press.

Corfield, T. F. (1973). Elephant mortality in the Tsavo National Park, Kenya. *East African Wildlife Journal* **11**, 339–68.

Crawley, M. J. (1983). *Herbivory: The Dynamics of Animal–Plant Interactions.* Oxford: Blackwell Scientific.

Croze, H. (1972). A modified photogrammetric technique for assessing age-structure of elephant populations and its uses in Kidepo National Park. *East African Wildlife Journal* **10**, 91–115.

Croze, H. (1974). The Seronera bull problem. II. The trees. *East African Wildlife Journal* **12**, 29–47.

Croze, H., Hillman, A. K. K. & Lang, E. M. (1981). Elephants and their habitats: How do they tolerate each other. In *Dynamics of Large Mammal Populations*, ed. C. W. Fowler & T. D. Smith, pp. 297–316. New York: John Wiley & Sons.

Deraniyagala, P. E. P. (1955). *Some Extinct Elephants, their Relatives, and the Two Living Species.* Colombo: National Museum of Ceylon.

Digby, S. (1971). *War Horse and Elephant in the Delhi Sultanate.* Oxford: Orient Monographs.

Dobias, R. (1985). Elephants in Thailand. *WWF Monthly Report,* December 1985, pp. 307–12. Gland, Switzerland: World Wildlife Fund.

Dobias, R. J. (1987). Elephants in Thailand: An overview of their status and conservation. *Tigerpaper* **14**, 19–24.

Dollard, J., Miller, N. E., Mowrer, O. H., Sears, G. H. & Sears, R. R. (1939). *Frustration and Aggression.* New Haven: Yale University Press.

Dougall, H. W., Drysdale, V. M. & Glover, P. E. (1964). The chemical composition of Kenya browse and pasture herbage. *East African Wildlife Journal* **2**, 86–121.

Douglas-Hamilton, I. (1972). On the ecology and behaviour of the African elephant. Unpublished D.Phil. thesis, University of Oxford.

Douglas-Hamilton, I. (1980). African elephant ivory trade study: Final report (excerpts). *Elephant* **1**, 69–99.

Douglas-Hamilton, I. (1987). African elephants: Population trends and their causes. *Oryx* **21**, 11–24.

Douglas-Hamilton, I. & Douglas-Hamilton, O. (1975). *Among the Elephants.* London: Collins & Harvill Press.

Ehrlich, P. R. & Birch, L. C. (1967). The 'Balance of Nature'. *American Naturalist* **101**, 97–107.

Edroma, E. L. (1981). The role of grazing in maintaining high-species composition in *Imperata* grassland in Rwenzori National Park, Uganda. *African Journal of Ecology* **19**, 215–33.

Edroma, E. L. (1984). Effects of burning and grazing on the productivity and number of plants in Queen Elizabeth National Park, Uganda. *African Journal of Ecology* **22**, 165–74.

Eisenberg, J. F., McKay, G. M. & Jainudeen, M. R. (1971). Reproductive behaviour of the Asiatic elephant (*Elephas maximus maximus* L.). *Behavior* **38**, 193–225.

Eisenberg, J. F. & Seidensticker, J. (1976). Ungulates in southern Asia: A consideration of biomass estimates for selected habitats. *Biological Conservation* **10**, 293–308.

Elder, W. H. & Rodgers, D. H. (1974). Immobilization and marking of African elephants and the prediction of body weight from foot circumference. *Mammalia* **38**, 33–53.

Eltringham, S. K. (1974). Changes in the large mammal community of Mweya Peninsula, Rwenzori National Park, Uganda, following removal of hippopotamus. *Journal of Applied Ecology* **11**, 855–65.

Eltringham, S. K. (1980). A quantitative assessment of range usage by large African mammals with particular reference to the effects of elephants on trees. *African Journal of Ecology* **18**, 53–71.

Eltringham, S. K. (1982). *Elephants.* Poole, Dorset: Blandford Press.

Field, C. R. (1971). Elephant ecology in the Queen Elizabeth National Park, Uganda. *East African Wildlife Journal* **9**, 99–123.

Field, C. R. (1976). Palatability factors and nutritive values of the food of buffaloes (*Synercus caffer*) in Uganda. *East African Wildlife Journal* **14**, 181–201.

Field, C. R. & Ross, I. C. (1976). The savanna ecology of Kidepo Valley National Park. II. Feeding ecology of elephant and giraffe. *East African Wildlife Journal* **14**, 1–15.

Fletcher, F. W. F. (1911). *Sport on the Nilgiris and in Wynaad.* London: Macmillan.

Fowler, C. W. (1981). Density dependence as related to life history strategy. *Ecology* **62**, 602–10.

Fowler, C. W. & Smith, T. (1973). Characterizing stable populations: An application to the African elephant population. *Journal of Wildlife Management* **37**, 513–23.

Francis, W. (1906). *Madras District Gazetteers*, vol. 1 (*South Arcot*). Madras: Government Press.

Frankel, O. H. & Soulé, M. E. (1981). *Conservation and Evolution.* Cambridge University Press.

Franklin, I. R. (1980). Evolutionary change in small populations. In *Conservation Biology: An Evolutionary – Ecological Perspective,* ed. M. E. Soulé & B. A. Wilcox, pp. 135–49. Sunderland, Massachusetts: Sinauer Associates.

Freeland, W. J. & Janzen, D. H. (1974). Strategies in herbivory by mammals: The role of plant secondary compounds. *American Naturalist* **108**, 269–89.

Freeman, D. (1980). *Elephants – The Vanishing Giants.* London: Hamlyn.

Gadgil, M. & Prasad, S. N. (1984). Ecological determinants of life history evolution of two Indian bamboo species. *Biotropica* **16**, 161–72.

Gilpin, M. E. & Soulé, M. E. (1986). Minimum viable populations: Processes of species extinction. In *Conservation Biology: The Science of Scarcity and Diversity,* ed. M. E. Soulé, pp. 19–34. Sunderland, Massachusetts: Sinauer Associates.

Gittins, S. P. & Akonda, A. W. (1982). What survives in Bangladesh? *Tigerpaper* **9**, 5–11.

Gooneratne, F. W. F. (1967). The Ceylon elephant *Elephas maximus zeylanicus* – its decimation and fight for survival. *Ceylon Journal of Historical and Social Studies* **10**, 149–60.

Guy, P. R. (1975). The daily food intake of the African elephant, *Loxodonta africana* Blumenbach, in Rhodesia. *Arnoldia Rhodesia* **7**, 1–8.

Guy, P. R. (1976). The feeding behaviour of elephant (*Loxodonta africana*) in the Sengwa area, Rhodesia. *South African Journal of Wildlife Research* **6**, 55–63.

Hairston, N. G., Smith, F. E. & Slobodkin, L. B. (1960). Community structure, population control, and competition. *American Naturalist* **94**, 421–5.

Hall-Martin, A. (1980). Elephant survivors. *Oryx* **15**, 355–62.

Hall-Martin, A. (1981). Conservation and management of elephants in the Kruger National Park, South Africa. In *The Status and Conservation of Africa's Elephants and Rhinos,* ed. D. H. M. Cumming & P. Jackson, pp. 104–15. Gland, Switzerland: International Union for Conservation of Nature and Natural Resources.

Hall-Martin, A. & Ruther, H. (1979). Application of stereo photogrammetric techniques for measuring African elephants. *Koedoe* **22**, 187–98.

Hanks, J. (1972a). Reproduction of elephant, *Loxodonta africana,* in the Luangwa Valley, Zambia. *Journal of Reproduction and Fertility* **30**, 13–26.

Hanks, J. (1972b). Growth of the African elephant (*Loxodonta africana*). *East African Wildlife Journal* **10**, 251–72.

Hanks, J. (1979). *A Struggle for Survival – the Elephant Problem.* Feltham, England: Country Life Books.

Hanks, J. (1981). Characterization of population condition. In *Dynamics of Large Mammal Populations,* ed. C. W. Fowler & T. D. Smith, pp. 47–73. New York: John Wiley & Sons.

Hanks, J. & McIntosh, J. E. A. (1973). Population dynamics of the African elephant (*Loxodonta africana*). *Journal of Zoology, London* **169**, 29–38.

Harthoorn, A. M. (1976). *The Chemical Capture of Animals.* London: Balliere Tindall.

Heath, B. R. & Field, C. R. (1974). Elephant endurance on Galana ranch, Kenya. *East African Wildlife Journal* **12**, 239–42.

Hicks, H. G. (1928). *Revised Working Plan for the Mudumalai Forests.* Madras: Government Press.

Hoffmann, T. W. (1975). Elephants in Sri Lanka, their number and distribution. *Loris* **13**, 278–80.

Hoffmann, T. W. (1978). Distribution of elephants in Sri Lanka. *Loris* **14**, 366–67.

Hooijer, D. A. (1972). Prehistoric evidence for *Elephas maximus* Linn. in Borneo. *Nature, London* **239**, 228.

Holling, C. S. (1973). Resilience and stability of ecological systems. *Annual Review of Ecology and Systematics* **4**, 1–23.

Imperial Gazetteer of India. (1907). *The Indian Empire,* New edition, Oxford.

INPUT (1983). *Felephence Manual.* Malaysia: Institut Pembangunan Tanah Felda.

Ishwaran, N. (1983). Elephants and woody-plant relationships in Gal Oya, Sri Lanka. *Biological Conservation* **26**, 255–70.

Ishwaran, N. & Banda, P. A. (1982). Conservation of the Sri Lankan elephant: Planning and management of the Wasgomuwa – Maduru Oya – Gal Oya complex of reserves. WWF Project 1783, final report. Gland, Switzerland: World Wildlife Fund.

Jachmann, H. (1980). Population dynamics of the elephants (*Loxodonta africana*) in Kasungu National Park, Malawi. *Netherland Journal of Zoology* **30**, 622–34.

Jachmann, H. & Bell, R. H. V. (1984). Why do elephants destroy woodland? *African Elephant and Rhino Group Newsletter,* June 1984, pp. 9–10.

Jackson, P., ed. (1982). *Elephants and Rhinos: Time for Decision.* Gland, Switzerland: International Union for Conservation of Nature and Natural Resources.

Jainudeen, M. R., Katongole, C. B. & Short, R. V. (1972). Plasma testosterone levels in relation to musth and sexual activity in the male Asiatic elephant, *Elephas maximus*. *Journal of Reproduction and Fertility* **29**, 99–103.

Jainudeen, M. R., McKay, G. M. & Eisenberg, J. F. (1972). Observations on musth in the domesticated Asiatic elephant (*Elephas maximus*). *Mammalia* **36**, 247–61.

Jennrich, R. I. & Turner, F. B. (1969). Measurement of non-circular home range. *Journal of Theoretical Biology* **22**, 227–37.

Jerdon, T. C. (1874). *The Mammals of India.* London: John Whelden.

Johnsingh, A. J. T. (1980). Ecology and behaviour of the dhole or Indian wild dog, *Cuon alpinus* Pallas 1811, with special reference to predator–prey relations at Bandipur. Ph.D. thesis, Madurai Kamaraj University.

Joseph, S. J. (1969). *Working Plan for the Coimbatore North Forest Division* (1970–71 to 1979–80). Madras: Office of the Chief Conservator of Forests.

Kala, J. C. (1979). *Working Plan for the Coimbatore North Forest Division* (1980–81 to 1989–90). Madras: Office of the Chief Conservator of Forests.

Kengle, R. P. (1972). *The Kautilya Arthasastra,* part 2. Bombay: University of Bombay.

Khan, M. A. R. (1980). On the distribution and population status of the Asian elephant in Bangladesh. In *The Status of the Asian Elephant in the Indian Sub-continent* (IUCN/SSC Report), ed. J. C. Daniel, pp. 63–72. Bombay: Bombay Natural History Society.

Khan, M. b. M. (1967). Movements of a herd of elephants in Upper Perak. *Malayan Nature Journal* **20**, 18–23.

Khan, M. b. M. (1985). Elephants in Peninsular Malaysia. *WWF Monthly Report,* December 1985, pp. 297–99. Gland, Switzerland: World Wildlife Fund.

Krishnan, M. (1972). An ecological survey of the larger mammals of peninsular India. The Indian elephant. *Journal of the Bombay Natural History Society* **69**, 297–315.

Kurt, F. (1974). Remarks on the social structure and ecology of the Ceylon elephant in the Yala National Park. In *The Behaviour of Ungulates and its Relation to Management*, ed. V. Geist & F. Walther, vol. 2, pp. 618–34. Morges, Switzerland: International Union for Conservation of Nature and Natural Resources.

Lahiri Choudhury, D. K. (1980). An interim report on the status and distribution of elephants in northeast India. In *The Status of the Asian Elephant in the Indian Sub-continent* (IUCN/SSC Report), ed. J. C. Daniel, pp. 43–58. Bombay: Bombay Natural History Society.

Lahiri Choudhury, D. K. (1986). Elephants in northeast India. *WWF Monthly Report*, January 1986, pp. 7–17. Gland, Switzerland: World Wildlife Fund.

Lamprey, H. F., Glover, P. E., Turner, M. I. M. & Bell, R. H. V. (1967). Invasion of the Serengeti National Park by elephants. *East African Wildlife Journal* **5**, 151–66.

Lang, E. M. (1980). Observations on growth and molar change in the African elephant. *African Journal of Ecology* **18**, 217–34.

Laws, R. M. (1966). Age criteria for the African elephant *Loxodonta africana*. *East African Wildlife Journal* **4**, 1–37.

Laws, R. M. (1969). The Tsavo Research Project. *Journal of Reproduction and Fertility* (Suppl.) **6**, 495–531.

Laws, R. M. (1970). Elephants as agents of habitat and landscape change in East Africa. *Oikos* **21**, 1–15.

Laws, R. M., Parker, I. S. C. & Johnstone, R. C. B. (1975). *Elephants and their Habitats*. Oxford: Clarendon Press.

Lekagul, B. & McNeely, J. A. (1977). Elephants in Thailand: Importance, status and conservation. *Tigerpaper* **4**, 22–5.

Leshner, A. I. (1978). *An Introduction to Behavioral Endocrinology*. New York: Oxford University Press.

Leuthold, W. (1977). Spatial organization and strategy of habitat utilization of elephants in Tsavo National Park, Kenya. *Zeitschrift für Säugetierkunde* **42**, 358–79.

Lewis, D. M. (1984). Demographic changes in the Luangwa valley elephants. *Biological Conservation* **29**, 7–14.

Lokemoen, J. T., Doty, H. A., Sharp, D. E. & Neaville, J. E. (1982). Electric fences to reduce mammalian predation on waterfowl nests. *Wildlife Society Bulletin* **10**, 318–23.

Lorenz, K. Z. (1966). *On Aggression*. London: Methuen.

MacArthur, R. H. & Wilson, E. O. (1967). *The Theory of Island Biogeography*. Princeton University Press.

Mace, G. M., Harvey, P. H. & Clutton-Brock, T. H. (1983). Vertebrate home-range size and energetic requirements. In *The Ecology of Animal Movement*, ed. I. R. Swingland & P. J. Greenwood, pp. 32–53. Oxford: Clarendon Press.

Martin, E. B. (1980). The craft, the trade and the elephants. *Oryx* **15**, 363–66.

McBee, R. H. (1971). Significance of intestinal microflora in herbivory. *Annual Review of Ecology and Systematics* **2**, 165–76.

McCullagh, K. G. (1969). The growth and nutrition of the African elephant. II. The chemical nature of the diet. *East African Wildlife Journal* **7**, 91–8.

McCullagh, K. G. (1973). Are African elephants deficient in essential fatty acids? *Nature, London* **242**, 267–68.

McKay, G. M. (1973). The ecology and behavior of the Asiatic elephant in southeastern Ceylon. *Smithsonian Contributions to Zoology* **125**, 1–113.

McNab, B. K. (1963). Bioenergetics and the determination of home range size. *American Naturalist* **97**, 133–40.

McNaughton, S. J. (1976). Serengeti migratory wildebeest: Facilitation of energy flow by grazing. *Science* **191**, 92–4.

McNaughton, S. J. (1979). Grassland–herbivore dynamics. In *Serengeti – Dynamics of an ecosystem*, ed. A. R. E. Sinclair & M. Norton-Griffiths, pp. 46–81. Chicago: University of Chicago Press.

McNeely, J. A. (1978). Management of elephants in Southeast Asia. In *Proceedings of the BIOTROP Symposium on Animal Populations and Wildlife Management in Southeast Asia*, pp. 219–25. Bogor, Indonesia: SEAMEO – BIOTROP.

Mishra, B. K. & Ramakrishnan, P. S. (1983a). Secondary succession subsequent to slash and burn agriculture at higher elevations of north-east India. I. Species diversity, biomass and litter production. *Acta Œcologica/Œcologia Applicata* **4**, 95–107.

Mishra, B. K. & Ramakrishnan, P. S. (1983b). Secondary succession subsequent to slash and burn agriculture at higher elevations of north-east India. II. Nutrient cycling. *Acta Œcologia/Œcologia Applicata* **4**, 237–45.

Mishra, B. K. & Ramakrishnan, P. S. (1983c). Slash and burn agriculture at higher elevations in north-eastern India. II. Soil fertility changes. *Agriculture, Ecosystems and Environment* **9**, 83–96.

Mishra, J. (1971). An assessment of annual damage to crops by elephants in Palamau District, Bihar. *Journal of the Bombay Natural History Society* **68**, 307–10.

Morris, R. C. (1958). Note on the use of bamboo gun rocket for scaring wild animals out of cultivation. *Journal of the Bombay Natural History Society* **55**, 344–5.

Moss, C. J. (1982). *Portraits in the Wild: Behavior Studies of East African Mammals*. Chicago: University of Chicago Press.

Mueller-Dombois, D. (1972). Crown distortion and elephant distribution in the woody vegetations of Ruhuna National Park, Ceylon. *Ecology* **53**, 208–27.

Murdoch, W. W. (1966). 'Community structure, population control and competition', a critique. *American Naturalist* **100**, 219–26.

Myers, N. (1973). Tsavo National Park, Kenya, and its elephants: An interim appraisal. *Biological Conservation* **5**, 123–32.

Nair, P. V. & Gadgil, M. (1978). The status and distribution of elephant populations of Karnataka. *Journal of the Bombay Natural History Society* **75** (Suppl.), 1000–16.

Nair, P. V., Sukumar, R. & Gadgil, M. (1980). The elephant in south India – A review. In *The Status of the Asian Elephant in the Indian Sub-continent* (IUCN/SSC Report), ed. J. C. Daniel, pp. 9–19. Bombay: Bombay Natural History Society.

Nash, S. V. & Nash, A. D. (1985). The status and ecology of the Sumatran elephant (*Elephas maximus sumatranus*) in the Padang Sugihan Wildlife Reserve, South Sumatra. Final Report, World Wildlife Fund Project 3133, Bogor.

Neginhal, S. G. (1974). *Project Tiger: Management Plan of the Bandipur Tiger Reserve*. Bangalore: Karnataka Forest Department.

Nicholson, F. A. (1898). *Madras District Manual, Coimbatore*, vol. 2. Madras: Government Press.

Norton-Griffiths, M. (1979). The influence of grazing, browsing, and fire on the vegetation dynamics of the Serengeti. In *Serengeti – Dynamics of an Ecosystem*, ed. A. R. E. Sinclair & M. Norton-Griffiths, pp. 310–52. Chicago: University of Chicago Press.

O'Brien, S. J., Roelke, M. E., Marker, L., Newman, A., Winkler, C. A., Meltzer, D., Colly, L., Evermann, J. F., Bush, M. & Wildt, D. E. (1985). Genetic basis for species vulnerability in the cheetah. *Science* **227**, 1428–34.

Olivier, R. (1978a). Distribution and status of the Asian elephant. *Oryx* **14**, 379–424.

Olivier, R. C. D. (1978b). On the ecology of the Asian elephant. Unpublished Ph.D. thesis, University of Cambridge.

Owen-Smith, N. (1982). Factors influencing the consumption of plant products by large herbivores. In *Ecological Studies,* vol. 42 (*Ecology of Tropical Savannas*), ed. B. J. Huntley & B. H. Walker, pp. 359–404. Berlin: Springer-Verlag.
Parker, I. S. C. & Martin, E. B. (1982). How many elephants are killed for the ivory trade? *Oryx* **16**, 235–39.
Peters, R. H. (1983). *The Ecological Implications of Body Size.* Cambridge University Press.
Phillipson, J. (1975). Rainfall, primary productivity and carrying capacity of Tsavo National Park (East), Kenya. *East African Wildlife Journal* **13**, 171–201.
Pielou, E. C. (1977). *Mathematical Ecology.* New York: Wiley–Interscience.
Pienaar, U. de V. (1969). Why elephant culling is necessary. *African Wildlife* **23**, 181–94.
Piesse, R. L. (1982). Wildlife-proof barriers in India. Field Document. Rome: FAO.
Prasad, K. (1981). African ivory trade in India. *Indian Forester,* June 1981, pp. 384–86.
Prasad, S. N. & Gadgil, M. (1981). Conservation and management of bamboo resources of Karnataka. Bangalore: Karnataka State Council for Science and Technology.
Prasad, S. N. & Sharatchandra, H. C. (1984). Primary production and consumption in the deciduous forest ecosystem of Bandipur in south India. *Proceedings of the Indian Academy of Sciences (Animal Sciences)* **93**, 83–97.
Puri, G. S. (1960). *Indian Forest Ecology.* (2 vols.) New Delhi: Oxford Book and Stationery Co.
Pyke, G. H. (1983). Animal movements: An optimal foraging approach. In *The Ecology of Animal Movement,* ed. I. R. Swingland & P. J. Greenwood, pp. 7–31. Oxford: Clarendon Press.
Pyke, G. H., Pulliam, H. R. & Charnov, E. L. (1977). Optimal foraging: A selective review of theory and test. *Quarterly Review of Biology* **52**, 137–53.
Ralls, K., Brugger, K. & Ballou, J. (1979). Inbreeding and juvenile mortality in small populations of ungulates. *Science* **206**, 1101–3.
Rao, H. S. (1957). History of our knowledge of the Indian fauna through the ages. *Journal of the Bombay Natural History Society* **54**, 251–80.
Rodgers, D. H. & Elder, W. H. (1977). Movements of elephants in Luangwa Valley, Zambia. *Journal of Wildlife Management* **41**, 56–62.
Rosenthal, G. A. & Janzen, D. H., eds (1979). *Herbivores: Their Interactions with Secondary Plant Metabolites.* New York: Academic Press.
Sadleir, R. M. F. S. (1969). *The Ecology of Reproduction in Wild and Domestic Mammals.* London: Methuen.
Salim Ali, M. (1927). The Moghul Emperors of India as naturalists and sportsmen. *Journal of the Bombay Natural History Society* **31**, 833–61.
Sanderson, G. P. (1878). *Thirteen Years among the Wild Beasts of India.* London: W. H. Allen.
Santiapillai, C. (1987). Action plan for Asian elephant conservation – a country by country analysis. Unpublished report of the IUCN/SSC Asian Elephant Specialist Group. Gland, Switzerland: World Wide Fund for Nature.
Santiapillai, C. & Suprahman, H. (1985). Elephants in Indonesia (Sumatra). *WWF Monthly Report,* December 1985, pp. 287–96. Gland, Switzerland: World Wildlife Fund.
Sayer, J. (1983). Nature conservation priorities in Laos. *Tigerpaper* **10**, 10–14.
Schaller, G. B. (1967). *The Deer and the Tiger.* Chicago: University of Chicago Press.
Schoener, T. W. (1971). Theory of feeding strategies. *Annual Review of Ecology and Systematics* **2**, 369–404.
Scott, J. P. (1958). *Aggression.* Chicago: University of Chicago Press.
Scott, J. P. (1962). Critical periods in behavioural development. *Science* **138**, 949–58.
Seidensticker, J. (1984). *Managing Elephant Depredations in Agricultural and Forestry Projects.* Washington, D.C.: The World Bank.

Shaffer, M. L. (1981). Minimum population sizes for species conservation. *BioScience* **31**, 131–34.

Shahi, S. P. (1980). Report of the Asian Elephant Specialist Group, Central India Task Force. In *The Status of the Asian Elephant in the Indian Sub-continent* (IUCN/SSC Report), ed. J. C. Daniel, pp. 35–42. Bombay: Bombay Natural History Society.

Shahi, S. P. & Chowdhury, S. (1986). Elephants in Central India. *WWF Monthly Report*, January 1986, pp. 19–22. Gland, Switzerland: World Wildlife Fund.

Shamasastry, R. (1960). *Kautilya's Arthasastra*. Mysore: Mysore Printing and Publishing House.

Short, J. (1981). Diet and feeding behaviour of the forest elephant. *Mammalia* **45**, 177–85.

Sikes, S. K. (1968). Habitat stress and arterial disease in elephants. *Oryx* **9**, 286–92.

Sikes, S. K. (1971). *The Natural History of the African Elephant*. London: Weidenfeld and Nicolson.

Sinclair, A. R. E. (1975). The resource limitation of trophic levels in tropical grassland ecosystems. *Journal of Animal Ecology* **44**, 497–520.

Sinclair, A. R. E. (1979). The eruption of the ruminants. In *Serengeti – Dynamics of an Ecosystem*, ed. A. R. E. Sinclair & M. Norton-Griffiths, pp. 82–103. Chicago: University of Chicago Press.

Sinclair, A. R. E. (1981). Environmental carrying capacity and the evidence for over abundance. In *Problems in Management of Locally Abundant Wild Mammals*, ed. P. A. Jewell & S. Holt, pp. 247–57. New York: Academic Press.

Sinclair, A. R. E. (1983). The function of distance movements in vertebrates. In *The Ecology of Animal Movement*, ed. I. R. Swingland & P. J. Greenwood, pp. 240–58. Oxford: Clarendon Press.

Sinclair, A. R. E. & Norton-Griffiths, M., eds (1979). *Serengeti – Dynamics of an Ecosystem*. Chicago: University of Chicago press.

Singh, J. S., Lauenroth, W. K. & Steinhorst, R. K. (1975). Review and assessment of various techniques for estimating net aerial primary production in grasslands from harvest data. *Botanical Review* **41**, 182–228.

Singh, V. B. (1978). The elephant in U.P. (India) – A resurvey of its status after 10 years. *Journal of the Bombay Natural History Society* **75**, 71–82.

Slobodkin, L. B., Smith, F. E. & Hairston, N. G. (1967). Regulation in terrestrial ecosystems, and the implied balance of nature. *American Naturalist* **101**, 109–24.

Smuts, G. L. (1977). Reproduction and population characteristics of elephants in the Kruger National Park. *Journal of the South African Wildlife Management Association* **5**, 1–10.

Stracey, P. D. (1963). *Elephant Gold*. London: Weidenfeld and Nicolson.

Sukumar, R. (1985). Ecology of the Asian elephant (*Elephas maximus*) and its interaction with man in south India. Unpublished Ph.D. thesis, Indian Institute of Science, Bangalore.

Sukumar, R. (1986*a*). The elephant populations of India – Strategies for conservation. *Proceedings of the Indian Academy of Sciences (Animal Sciences/Plant Sciences) (Supplement)*, November 1986, pp. 59–71.

Sukumar, R. (1986*b*). Elephant–man conflict in Karnataka. In *Karnataka – State of Environment Report 1984–85*, ed. C. J. Saldanha, pp. 46–58. Bangalore: Centre for Taxonomic Studies.

Sukumar, R. (1989*a*). Ecology of the Asian elephant in southern India. I. Movement and habitat utilization patterns. *Journal of Tropical Ecology* **5**, 1–18.

Sukumar, R. (1989*b*). Ecology of the Asian elephant in southern India. II. Feeding habits and crop raiding patterns. *Journal of Tropical Ecology* **5**. (In the press.)

Sukumar, R., Bhattacharya, S. K. & Krishnamurthy, R. V. (1987). Carbon isotopic evidence for different feeding patterns in an Asian elephant population. *Current Science* **56**, 11–14.

Sukumar, R. & Gadgil, M. (1988). Male–female differences in foraging on crops by Asian elephants. *Animal Behaviour* **36**, 1233–1235.

Sukumar, R., Joshi, N. V. & Krishnamurthy, V. (1988). Growth in the Asian elephant. *Proceedings of the Indian Academy of Sciences (Animal Sciences)* **97**, 561–71.

Templeton, A. R. (1986). Coadaptation and outbreeding depression. In *Conservation Biology: The Science of Scarcity and Diversity*, ed. M. E. Soulé, pp. 105–16. Sunderland, Massachusetts: Sinauer Associates.

Terborgh, J. W. (1976). Island biogeography and conservation: Strategy and limitations. *Science* **193**, 1029–30.

Trautmann, T. R. (1982). Elephants and the Mauryas. In *India: History and Thought*, ed. S. N. Mukherjee, pp. 254–81. Calcutta: Subarnarekha.

Ulrich, R. E. (1966). Pain as a cause of aggression. *American Zoologist* **6**, 643–62.

Vancuylenberg, B. W. B. (1977). Feeding behaviour of the Asiatic elephant in southeast Sri Lanka in relation to conservation. *Biological Conservation* **12**, 33–54.

Varadarajaiyer, E. S. (1945). *The Elephant in the Tamil Land*. Annamalainagar, Tamilnadu: Annamalai University.

Vesey-Fitzgerald, D. (1973). Browse production and utilization in Tarangire National Park. *East African Wildlife Journal* **11**, 291–305.

Vijayan, V. S. (1980). Status of elephants in Periyar Tiger Reserve. In *The Status of the Asian elephant in the Indian Sub-continent* (IUCN/SSC Report), ed. J. C. Daniel, pp. 31–4. Bombay: Bombay Natural History Society.

von Bertalanffy, L. (1938). A quantitative theory of organic growth. *Human Biology* **10**, 181–213.

von Lengerke, H. J. (1977). *The Nilgiri – Weather and Climate of a Mountain Area in South India*. Wiesbaden: Franz Steiner Verlag.

Vo Quy (1986). Vietnam: Coming back from environment destruction. *IUCN Bulletin* **17**, 59.

Walker, C. (1982). Desert dwellers. *Elephant* **2**, 135–39.

Warmington, E. H. (1974). *The Commerce between the Roman Empire and India*. Delhi: Vikas Publishing House.

Wealth of India. (1948). Vol. 1. New Delhi: Council for Scientific and Industrial Research.

Weir, J. S. (1973). Exploitation of water soluble soil sodium by elephants in Murchison Falls National Park, Uganda. *East African Wildlife Journal* **11**, 1–7.

Western, D. & van Praet, C. (1973). Cyclical changes in the habitat and climate of an East African ecosystem. *Nature, London* **241**, 104–6.

Westoby, M. (1974). Analysis of diet selection by large generalist herbivores. *American Naturalist* **108**, 290–304.

Whittaker, R. H. (1970). *Communities and Ecosystems*. London: Collier-Macmillan.

Wilcox, B. A. (1980). Insular ecology and conservation. In *Conservation Biology: An Evolutionary–Ecological perspective*, ed. M. E. Soulé & B. A. Wilcox, pp. 95–117. Sunderland, Massachusetts: Sinauer Associates.

Williams, J. H. (1950). *Elephant Bill*. London: Rupert Hart-Davis.

Wing, L. D. & Buss, I. O. (1970). Elephants and Forests. *Wildlife Monographs*, no. 19.

Wu, L. S.-Y. & Botkin, D. B. (1980). Of elephants and men: A discrete, stochastic model for long-lived species with complex life histories. *American Naturalist* **116**, 831–49.

Wyatt, J. R. & Eltringham, S. K. (1974). The daily activity of the elephant in the Rwenzori National Park, Uganda. *East African Wildlife Journal* **12**, 273–89.

INDEX

Page numbers in italics refer to illustrations and figures.

Acacia leucophloea 93, *94*, 99
Acacia sinuata 149
Acacia spp. 32, 71
   bark collection 149, 150
   browse diet 74
   damage to 57
   plantations 154, 155
   protein levels 83
   spines 84
   tannin content 84
Acacia suma 93, 95–6, 99
Accelerated Mahaweli Development
   Programme: *see* Mahaweli Ganga
   Development Project
acepromazine 217
Adansonia digitata 100
African elephant 2
   aggression 141–2
   browsing 85
   chemical immobilization 216
   demographic parameters 199–200
   home range size 64–5
   impact on vegetation 86
   number of plants in diet 81
   population models 174
   population studies 105
   seasonal distribution 68
Agasthyamalai *12*, 15, *30*
age
   at death 185
   distribution 186–7, 191–3
   field methods of estimating 54–5, 224, 226
   and growth parameters *225*
aggressive behaviour 136, 137, 140–2
agriculture 2
   expansion in Indonesia 34
   habitat loss to 144
   permanent cultivation 33–4
   planning 218
   shifting cultivation 32–3, 144–5, 151, 208–9
Ailanthus excelsa 154

Ain-i-Akbari 5
Akkurjorai Reserve 132
alcohol 79–80
Alexander the Great 3
alkaloids 84
altitude 11
American bombing of elephants 8, 37
Amparai Sanctuary 26
Anamalai *12*, 14, *30*, 34
anatomy 81
anchored *mela-shikar* 215
Andaman Islands 26, *31*
Andhra 4
Andhra Pradesh 13
Anogeissus latifolia 152, *153*, 209
anthrax 150, 182, 201
   transmission from livestock 209
Antilope cervicapra 52
Araikadavu stream 42
   water availability 60, 65
archeological remains 2
armies, elephant 3, 5
Arthasastra 3
Artocarpus integrifolia 108, 114
Arunachal 17
Arunachal Pradesh *30*
   sex ratio of population 204
Aryan people 2, 3
Ashambu hills 15
Asian Elephant Specialist Group of IUCN 10
   Central India Task Force 15
   Northeast India Task Force 17
Assam 17, *30*
Aswathamma 3
Attapadi hills *12*, 14
Australian wattle 154
Axis axis 51–2
Azadirachta indica 149, 154

Badaga tribe 54
bamboo 70

Index

bamboo (*cont.*)
  browse species 75
  elephants' feeding habits 149
  extraction 147–9, 209
  feeding on 71
  gun rockets 125
  invading after timber felling 146
  plantations 154
Bambusa arundinacea 70, *71*
  extraction 147–9
  plantations 154
banana 108
  damage by elephants 111, 114, 124
Bandipur *12*, 13, *30*
  ecological biomass 90
Bandipur National Park 159
Bangalore city 13
Bangladesh 17, 18, *30*
baobab 100
bark 43
  consumption 71, 72, 83–4
  forest produce collection 149, 150
  plantation trees 155
  protein level 85
  stripping from Malvales 93
  tannin levels 84
barking deer: *see Muntiacus muntjac*
barriers 213–14
Bejjaluhatti 116
Bengal 17, *30*
Betta Kurubu sect 54
Bhadra *30*
Bhadra Wildlife Sanctuary 12
Bhutan 17, *30*
Bia National Park (Ghana) 78
Bihar 15, *30*
Biligirirangan hills 13, 14
  movement patterns in 66
  study area 39, 40, 42
biomass
  grass 87, 88
  large mammalian herbivore 89–91
birth rate 179–82
blackbuck 52
bodyweight loss 82
bonnet monkey 52
Borneo 28, *29*, *31*
  permanent cultivation 33–4
Bos gaurus 51, *156*
Brahmaputra river 17
  area south of 17–18
breeding
  in captivity 6
  programmes 210–11
bridges 207
browse
  nutritive value 85
  plants in diet 67, 70–1, 74–8
  proportion in diet 71, 73–4
browsing
  bamboo 149
  habits 70–1

preference in dry season 83
species taken 70–1
time spent 73
Bubalus bubalis 52
buffalo 43, 52
buffer zones 214–15
bulls
  age distribution 185
  age frequency poached 168–9
  ageing 226
  aggressive behaviour 136, 137, 140–2
  captive 210
  crop raiding 111, 114–16, 131
  damage to buildings 137, *138*
  frequency of unit group sizes 119
  group size in crop raiding 116, *117*
  groups 51
  home range size *63*, 64
  hunting and effects on adult sex ratio 195
  life-tables *184*
  predicted trends in population *198*
  quantity of crop damage 123
  quantity of crops in diet 120–2
  removal of habitual crop raiders 215
  reproductive success 134
  sexual maturity 179
  tuskless 165, *167*, 177, 204
  use of bridges 207
Burma 7, 17, 18, 20, *31*
  capture in 8
  hunting mortality 36

C3 and C4 plants 85
calcium
  availability 84
  in bark 83–4
  content of finger millet 134
  requirements 82
calves
  mortality 200
  trends in population *192*
calving
  age 177–9, 188
  interval 105, 179–82, 188, *196*, 197
canals 161–2
Canis aureus 52
captive elephants 210
capture 2, 3, 35
  for domestication 107, 215
  nineteenth and twentieth centuries 8
  traditional methods 217
captures 9
Careya arborea 71
carrying capacity 86
  density-dependent change 188
  inter-specific competition 97–9
  resource limitation 97–9
  time lag 101
Cassia spp. 149
cattle 43; *see also* livestock
Cauvery river 13, 40

Index

cave paintings 2
cellulose digestion 81
cereals 57
Cervus unicolor 52
Chandragupta 3
Chap-Chur syringe dart 216
charcoal 147
charging 137, *138*
chemical
   immobilization 26
   repellants 215
Chikkahalli
   crop raiding 116
   herd movements 131
China 7, 18, 22, *31*
chital: *see* spotted deer
Chitawan 90
Chittagong hills 18
Chromolaena adenophora 145
Chromolaena odorata 144, 155
Chromolaena spp. 208
   bamboo suppression 148
clans 51
   movement 62–3
climate of study area 42–3
cobalt 84
coconut 108
   crop damage 58
   damage by elephants 111, *113*, *114*, 124
   parts eaten 133
Cocos nucifera: *see* coconut
coffee plantations 43, *49*
Coimbatore plains 40
colic 83
common langur 52
compensation for crop damage 218
compression hypothesis 99–100
conception rate 181, *182*
conservation
   biological diversity 202
   genetics 203, 204
   human interests 202
   minimum habitat size 206
   social security schemes 218
Convention on International Trade in Endangered Species of Wild Fauna and Flora (CITES) 171
corridors 206
crops
   consumed by elephants 111–14
   cultivated 108, 111
   damage 7, 57–8, 123–5
   insurance schemes 218
   killing of elephants in defence 168
   land under cultivation *110*
   methods of protection 211–16
   nutritive value 133–4
   palatability 133–4
   plantation 108, 111
   staple food 111
   see also raiding crops

culling 87, 100, 215
   Asian elephants 107
   effect on population dynamics 106
cultural associations 7
Cuon alpinus 52
Cymbopogon flexuosus 69
   palatability 133
   protein level 83
Cymbopogon spp. *49*, 149
Cyperaceae 69
cyprenorphine hydrochloride 216
Cyrus the Great 3

Dalbergia latifolia 146
Dalbhum tract 15
dams
   crop protection 213
   habitat effects 159, 206
   impact 34, 35
Dandeli Wildlife Sanctuary 11
Dangrek mountains 21
deciduous forest
   elephant density 60, 62
   habitat types 208
   herbivore biomass 90
   regions of 11, 12, 15, 17, 18
   study area 40, 44, *45*, 48
   use in wet season 66
Delhi Sultanate 5
demographic
   parameters 199–200
   vigour 197, 200
Dendrocalamus strictus 70, 147–9
density 60, 62
   dependence 101
   estimated seasonal 219–23
   estimates 54, 55–6
density-dependent negative feedback 188
deserts, African 100
diet
   browse and grass mix 85
   browse plants 74–8
   diversity 84
   number of plants 81
   optimal 85
digestion efficiency 81
disease 150
   effect on populations 207–8
   environmental stochasticity 203
   transmission from livestock 209
distribution 1, 60–2
   altitude 11
   countries 10
   historical 2, 4, 5
   Indian sub-continent 11–18
   regions 10
Dolichos biflorus 111
domestic elephants 210
   Burma 20
   chasing crop raiders 215
   handling captured elephants 217

domestication 2
  capture for 107
  nineteenth and twentieth centuries 8
Dravidian people 2
drinking water: *see* water
drought 203
dry season 43
  bodyweight loss 82
  browse preference 83
  distribution 60, 62
  mortality 200
  nutritional level of food plants 83
  quantity of forage consumed 78, 79
Dudhwa National Park 16

East Kalimantan 28, *29*
Eastern Ghats 6, 11, *12*, 13–14, *30*, *47*
  crop raiding 132
  sex ratio of population 204
  study area 39
eating habits 56
ecosystem dynamics 99–107
Elaeis guineensis
  crop raiding 126, 128–30
  parts eaten 133
electric fence 37, 126, 168, 169, 211–13
  costs 214
Elemalai 14–15
Elephant Control Scheme of the Forest Department (Burma) 35
Elephant Preservation Act (1873) 7
elephant–tree interaction 107
elephant–vegetation
  dynamics *103*, 104–6
  interaction 99–107
Elephas maximus 1, 2
  density in study area 50
  population decline 8
Eleusine coracana 111
  crop raiding 116, 119, 120
  damage area 123
  damage by elephants *112*
  mineral content 134
  palatability 133
  potential crop loss 125
  protein content 134
Endau–Rompin National Park 24
environment
  fluctuating 104
  and movement patterns 65
etorphine hydrochloride 216, 217
Eucalyptus spp. 32, 71
  damage to 57, 97
  plantations 43, *45*, *49*, 154, 155
Euphorbiaceae browse species 74, *76*, 78
evergreen forest 208
  regions of 11, 13, 14, 15, 17, 18
  study area 40, 44, *49*
evolutionary history 2
exchange, artificial 204–5

family groups
  age structure 55, 175–7, 180
  movement 62–3
  sex structure 175, 177
  size 51
family unit 50–1
Fazl, Abul 5
Federal Land Development Authority (Malaysia) 33, 126, *129*
feeding
  behaviour 97–8
  habits 70–1
  observations 55–6
  patterns 69
  rate 78
felling 208; *see also* logging
females
  ageing 226
  calcium requirements 82
  knowledge of migration routes 66
  life-table *186*
  mortality 186, 197
fermentation
  cellulose digestion 81
  fruit 80
fertility
  age distribution 193
  effect of adult sex ratio 195, 197
  and inbreeding 203
  levels 101
  in population dynamics 177–82
  population growth 189
  in population model 174–5
  schedule 188
  of study population 181
Ficus 71
finger millet 111
  crop raiding 116, 119, 120
  damage area 123
  damage by elephants *112*
  mineral content 134
  palatability 133
  potential crop loss 125
  protein content 134
finger of trunk 81
fire
  bamboo regeneration 148
  control 154
  crackers 124, 125
  effect on grasses 97
  effect on woody plants 99
  grass productivity 89
  habitat effects 151–4
  habitat management 209
  population dynamics effects 106
flashlights 126, 215
Flood Plains National Park 26
foetal development 200
food
  availability and home range size 65
  consumption 56

food (*cont.*)
　　effect of fire on availability 151, 152
　　nutritive value 82–4, *227–9*
　　plant analysis 58, *227–9*
　　population regulation 86
　　requirement 81
　　resource production and consumption 87
foot-and-mouth disease 150
forage 78–9
foraging
　　crop fields 120
　　nutrition and 80–5
forest
　　bamboo 18, 20
　　conversion to plantations 32
　　dry 15
　　effect of logging 146
　　grove 13
　　minor produce 149–50
　　natural and woody food plants 155
　　plantations 154–5, *156*
　　riparian 208
　　sal 15
　　semi-evergreen 14, 15, 18
　　study area 40
　　teak 20; *see also* teak
　　thorn 44, *45*, *47*
　　see also deciduous forest, evergreen forest, *shola* forest, tall grass forest
fruits 43, 149
　　eating 71
　　fermentation 80
　　food in rain forests 78
fuel 43, 146, 147

Gaddesal
　　crop raiding 116
　　potential crop loss 125
Gajasastra 2
Gal Oya National Park 26, 74
gall-nut 149
Ganesha worship 4
Garo hills 204
gaur 51–2, *156*
gene complex, specific coadapted 205
genetic
　　drift 203, 204
　　fitness 205
　　variation and fitness 203
　　viability of population 203
genotype 104
gestation 180
Ghaznavid kingdom 5
gingelly 108
Gir Sanctuary 98
glycosides, cyanogenic 84
god, elephant 4
gooseberry 149
Gramineae 69
　　elephants' preference for 133
　　hydrogen cyanide levels 87

grass
　　biomass 87, *88*
　　consumption 56–7
　　cultivated 133
　　extraction 149
　　invading after timber felling 146
　　nutritional levels 85, 97
　　in optimal diet 85
　　primary production 56, 87–9
　　production and consumption 90–1
　　proportion in diet 71, 73–4
　　protein level 85
　　rainfall in primary production 87, 89
　　regeneration after fire 151
　　swamp 85
　　tussock 82
grassland 13, 14, 20
　　elephant density 60, 62
　　study area 40, *45*, *47*, *49*
　　swampy 17, 67, 208
grazing
　　grass productivity 89
　　habits 70
　　species taken 69–70
　　time spent *73*
　　wet season 73
Grevillea robusta 32
　　plantations 43, 154, 155
Grewia spp. 71
Grewia tiliaefolia 57, 146
　　elephants' impact 91–3
growth
　　equations 54–5
　　relationships 224–6
Guizotia abyssinica 111
guns 125–6, 215
Gunung Leuser National Park 28

habitat 11
　　alteration by livestock grazing 150
　　artificial management 209
　　degradation 133
　　development in 7
　　diversity of types 208
　　effect of forest plantations 154–5
　　exploitation 32
　　fire effects 151–4
　　fragmentation 6, 132–3, 143–4, 206
　　human use 208
　　impact of elephants 87, 99
　　integrity 205–8
　　legal protection 37
　　livestock and quality 150
　　loss to dams 159
　　management and conservation 202
　　manipulation 209
　　minimum viable area 205–8
　　quality maintenance 150, 208–9
　　reduction 7, 8, 132–3, 143–4, 206
　　seasonal use 66–7
　　secondary 157–9

habitat (*cont.*)
  zones as sampling units 54
  zones of study area *46*, 47
Hasanur *41*
  crop raiding 120
  damage area 123
  herd movements 131
  potential crop loss 125
  rainfall 42
heat stress 153
height 55
  measuring 224
Helicteres isora 71
herbivore
  biomass 56–7, 89–91
  trophic level 86
herds
  crop raiding 114–16
  frequency of unit group sizes 119
  quantity of crops consumed *121*, 122–3
  size in crop raiding 116, *118*
Hevea brasiliensis
  crop raiding 126, *129*, 130
  damage by elephants 97
Himalayan foothills 16, 17
Himchari National Park 18
Himidurawa Tank region 74
history 2–8
hole digging 79, *80*
home range size 63–5
honey 43
horse gram 111
human
  impact 32
  interests, protection of 211–18
  interference in population density 106
  predators 2
  settlement 158
hunters 2
  ivory poaching 170
hunting 7, 35
  bulls and effects on adult sex ratio 195
  in Burma 20
  harassment and aggression 141
  numbers killed *9*
  seasonal 15
  see also poaching
Hurulu Reserve Forest 26
Hyaena hyaena 52
hydro-electric projects 11, 14, 15, 159
  effect on habitat 34–5
hydrogen cyanide 84
hydrology of study area 40, 42

identification of study elephants 54
illegal trade 35
immobilization, chemical 216–17
Immobilon 217
Imperata cylindrica 145
  effect of fires 152
  protein level 85
inbreeding depression 203

India
  central 15, *30*
  distribution 6
  historic distribution 2
  northern 16–18, *30*
  southern 11–15, *30*
  surveys 10
  wild population 6
Indian sub-continent 11–18
Indo-Gangetic plain 4
Indonesia 34
Indus valley civilization 2
inter-specific aggression 140
inter-specific competition 97, 98–9
  animal number regulation 98
International Union for Conservation of Nature
    and Natural Resources 10
intoxication 80
iron ore 15
irrigation 159
  dams 15
  habitat effects of schemes 34–5
  Tunga and Bhadra rivers project 11–12
Irula tribe 53
ivorine 210
ivory 2, 4, 35
  African 4
  carving 4, 170
  craftsmen 170, 173
  export of finished 171, 173
  importation 170–1
  price 171, *172*
  substitutes 210
  trade 4, 210
  value *172*, 173
ivory poaching 165, 166, 168
  organization 169–70
  trade 170–3

jackal 52
jackfruit 108, 114
Java 28
  transmigration scheme 34
Jehangir 6
Jenu Kuruba sect 54

Kabbini reservoir 13
Kadu Kuruba tribe 53–4
Kakankote 13
Kalinadi hydroelectric project 11
Kampuchea 22, *31*
Kanara *30*
Karnataka 11, 13
  damage to millet crops 36
Kattepura Reserve 132
Kautilya 3
Kaziranga hills
  herbivore biomass level 90
  sex ratio of population 204
Kerala North Wynad 13
Kerala South Wynad Sanctuary 13

Khasi hills 204
kheddah 3, 217, 218
  captures 8
kicking 137
killing 9
  methods 137, *139*, 140
kin groups 51
Kiziranga National Park 17
Kolipalya
  crop raiding 116, 120
  damage area 123
  herd movements 131, 132
  potential crop loss 125
Kruger National Park 100
Kydia calycina 57, 71
  elephants' impact 91–3
  regeneration after fire 152

lactation 82
Lahugala Trunk 26
Lake Manyara National Park (Tanzania) 64, 93
  calf mortality 200
Lambadi tribe 54
land-use
  pattern of study area 43
  planning 206
langur 52
Lantana camara 144, 155, 208
  bamboo suppression 148
Laos 22, *31*
large mammalian herbivore biomass 89–91
latex 84
legal protection 37
Leguminosae 69
  browse diet 74, *75*
  elephants' preference for 133
  hydrogen cyanide levels 87
  protein levels 83, 85
  woody species *76*
leopard 52
Leslie matrix model 175
lichen 149
life-tables 184, 185–7
light and grass productivity 89
Limonia acidissima 71, 149, 150
Lingayat community 54
livestock
  bamboo grazing 148
  disease 150
  domestic 43
  fodder 149
  grazing 57, 148, 150
  habitat competition 209
  inter-specific competition 98
logging 7, 32
  habitat effects 208
  industry 202
  use of elephants 35
loudspeaker 126
Lower Bhavani Reservoir 42
  blackbuck 52

Loxodonta africana 28; *see also* African elephant
Luangwa Valley 100, 101

Macaca radiata 52
Maduru Oya National Park 26
Mahabharatha 3
Mahaweli Ganga Development Project 24, 25, 26, 162–4
  dam construction 35
  translocation of elephants 217
Mainimukh Wildlife Sanctuary 18
maize 108
  damage area 123
  potential crop loss 125
makhnas: *see* bulls, tuskless
Malaysia 22–4, *31*
  electric fencing 213
  grass preference 85
  home range size 65
  palm species in diet 74
  permanent cultivation 33–4
  seasonal movement 63
male dispersal 51
Malnad *30*
  plateau 11–12
Malvales 69
  bark stripping 93
  browse diet 74, *75*
  protein levels 83
mammoth 2
man: *see* human
Manas Tiger Reserve 17
Mangifera indica 108, 149, 150
mango 108, 149, 150
Manipur *30*
manslaughter 36, 37, 135–42
  causes of aggression 140–2
  compensation 218
  injured elephants 140–1
  records 58, 135–6
mastodon 2
Mauryan kingdom 3
Mavallam 116
Mavanattam
  crop raiding 116
  herd movements 131
meat 3, 4, 35
  hunting for in Burma 20
  poaching 166
Meghalaya 17, *30*
  elephant population 204
  shifting cultivation 33
Mekong basin 22
mela-shikar 217
Melursus ursinus 52
menopause 188
migration routes 66
millet 111
  damage area 123
  potential crop loss 125
  quantity consumed in raids 57, 58

Index

minerals 82–4, 134
  content of food plants 229
mining 11, 15
Minneriya-Giritale Nature Reserve 26
Mizoram
  hunting for meat 35–6
  political unrest 37
Moeritherium 2
Moghuls 5, 6
monsoon crops 11, 108
morphological features 55
mortality 59
  adult sex ratio 193–4
  calf 200
  and calving interval 197
  causes of 166–8
  due to man 182, 184
  and inbreeding 203
  instantaneous rate 186–7
  juvenile 184, 190, 200
  Lake Manyara 200
  natural causes 182, *183*, 184
  numbers found dead 183–4
  patterns 188, *189*
  poaching 201
  population growth 189
  in population model 174–5
  rate 183, 193–4
  schedules *196*
  seasonal 200
moss 149
movement
  clans 62–3
  effect of development projects 161
  family 62–3
  foraging 67–8
  long-term 65–6
  pattern and inter-annual differences 65
  seasonal and crop raiding 131
  water availability 67–8
Moyar river valley 42, *47*
  dry season distribution 60
  study area 39, 40
  wet season distribution 62
Mudumalai Wildlife Sanctuary 13
  habitat use 67
Mudumalai–Bandipur tract 39
Muntiacus muntjac 52
Murchison Falls Park (Uganda) 193
Musa paradisiaca 108
  damage by elephants 111, 114, 124
musth 139, 140
mutation 203

Nagaland *30*
  hills 17, 204
  hunting for meat 35–6
  political unrest 37
  sex ratio of population 204
  shifting cultivation 33

Nagarhole *12*, 13, *30*
  herbivore density 155, *156*
  sex ratio of population 204
Nam Nao National Park 21
Namdapha National Park 17
Nanagunhe Nature Reserve 18
natural selection 203
neem
  fruit 149
  plantations 154
Nelliampathi *12*, 14
Nepal 16, *30*
Neydalapuram
  crop raiding 116
  herd movements 131
niger 111
Nilambur *12*, 14
Nilgala Jungle Corridor 26
Nilgiris *12*, 13, 14, *30*, 40
  electric fencing *212*, 213
  habitat use 66–7
  sex ratio of population 204
Nirdurgi river 42
nitrogen excretion 82
noosing 217
North Kanara 11, *12*
North Wynad 13
numbers, estimates 10
nutrition 80–5
nutritive value of food plants 82–4, *227–9*

oil drilling 32
oil palm
  crop raiding 126, 128–30
  destruction 36
  parts eaten 133
  plantations 34
Orissa 4, 5, 15, *30*
  shifting cultivation 33
Oryza sativa 111
  damage area 123
overgrazing 98

Pablakhali Wildlife Sanctuary 18
Padang Sugihan Reserve 27, 28
paddy 111
  damage area 123
Palakapya 2
Palamau Tiger Reserve 15
Palani hills 14
palatability 82–4
Palmae 69
  browse species 75
  eating 71
  elephants' preference for 133
Pandanus 71
Panicum miliare 111
  damage area 123
  potential crop loss 125
Panthera pardus 52

248   Index

Panthera tigris 52
paper industry 149
Parambikulam Sanctuary 14
Parambikulam–Aliyar Project 160–2
patrols 215
people
   encounters with elephants 137, 140
   killed 135–6
   methods of killing 137, *139*, 140
   see also human
Periyar *12*, 14–15, *30*
   reservoir 35
Periyar Tiger Reserve 15, *207*
   poaching 168
Petchabun mountains 21
Phoenix humilis 48, 71, 149, 150
photography 54, 55
   measuring height 224
Phyllanthus emblica 149
physiology 81–2
Piduang Sanctuary 18
pil-khana 4, 5
Pinus spp. 154
pipelines 161, 207
   barrier to cultivated areas 213
plant
   biomass 97
   C3 and C4 85
   chemical defences 84
   diversity after fires 152
   food 82–4, *227–9*
   nutritional level 82–4, 97, *227–9*
   physical defences 84
   product extraction 145–50
   production 97
   productivity and elephant density 106
   resources 32
   secondary compounds 84, 97, 133
   species 69
plant–herbivore interactions 102
plantations 34
   availability of food plants 155
   crops 111, 118
   forest and effect on habitat 154–5
   monoculture 32, 146, 155, 208
   tree damage by elephants 97
poaching 35, 36, 165, 166, 168
   and adult sex ratio 194, 195
   age frequency 168–9
   effective population size 204
   mortality rate 168, 184, 201
   reduction 210
   value of trade 173
   see also ivory
political situation 32, 37
   Laos, Kampuchea and Vietnam 22
   northeastern India 18
ponds 132, 159
population
   adults *192*
   and age distribution 191–3

aspects of dynamics relevant to southern India 175
calves *192*, 193
compression 144
cycles 199
decline 8
decline in North Kanara 11
demographic condition 197, 199–201
density and secondary vegetation 157–8
dynamics and demographic vigour 179
effective size 203, 204
estimates 29, *30–1*
genetic fitness 205
genetic viability 203
growth and calving interval *196*
growth trends 189–91
long-term fitness 203
minimum viable size 202–5
model validity 197
modelling 174, 188–201
natural regulation 209
physiological condition 197
predicted trends in adult males *198*
regulation 86–7
small isolated 205
specific coadapted gene complex 205
porcupine 128, 130
Porus, elephant army of 3
predation 142
predators 2
pregnancy 82
Presbytis entellus 52
primary production 86
   grass 56, 87–9
Proboscidae evolutionary history 2
protein
   content of food plants 85, *227–8*
   level of browse plants 67
   level of tall grass 67–8
   positive selection for high levels 83
   requirements 81–2
Pterocarpus marsupium 146
Pudukuyyanur
   crop raiding 116
   herd movements 131
pulp
   manufacture 149
   wood 43
Punjur
   crop raiding 120
   herd movements 131, 132

Queen Elizabeth National Park (Uganda) 193

radio-tracking, home range 64
ragi: *see* finger millet
raiding crops 2, 36–7, 57
   causes 130–4
   corridors 206
   deterrents 125–6
   economic loss 123–5

Index

raiding crops (*cont.*)
  frequency 114–16
  group sizes 116–19
  habitat destruction 132–3
  habitat reduction 206
  killing elephants 169
  porcupine 128, 130
  prevention 211–16
  quantity consumed 120–3
  quantity damaged 123–5
  rain forest habitat 126, 128–30
  reproductive success 134
  seasonality 114, 116
  wild pig 125, 128
rain forest 103
  carrying capacity 105
  crop raiding 126, 128–30
  food availability 157
  fruits as food 78
  primary versus secondary 157
rainfall 11
  and conception rate 181, *182*
  ecosystem stability-resilience 102–4
  effect on ecosystem 102
  movement patterns 65
  in primary production 87, 89
  seasonal use of habitat types 67
  of study area 42
  variation 102
rainy season: *see* wet season
Ramganga reservoir 16
record collection 58
registration of elephants 54
religious associations 7
reproduction
  in population dynamics 177–82
  rate in fluctuating environment 104
reproductive success 134
Reserved Forest status 38
reservoirs
  creation in elephant habitat 159–64
  habitat effects 144, 206–7
resilience, ecological 102
resource limitation 97
Revivon 217
rice: *see* paddy
Rig Veda 2
rinderpest 150, 209
Rishikesh–Chilla power channel 16
road construction 32
rogue elephants 37, 136, 216
Rompada 2
Rompun 217
rosewood 146
rubber
  crop raiding 126, *129*, 130
  damage by elephants 36, 97
  plantations 15
Ruhuna National Park 26, 74
Sabah 28, *29*
  agricultural expansion 33

Saccharum officinarum 108
  damage by elephants 114
sal 15
sambar 52, 57
sanctuaries 3
sandalwood 43
  felling 146
  plantations 154
  poaching 170
Sanderson, G. P. 8
Sansevieria 71
Santalum album 43, 154, 170
  felling 146
Sapindus emarginatus 149
Satkosia Gorge Sanctuary 15
savanna woodland 102–3
  carrying capacity 105
scrub vegetation 13, 14, 208
seasonal breeding 195
seasons 43
Sellucus Nikator 3
Serengeti ecosystem 86, 97
  inter-specific competition 98
  oscillation 101–2
Serengeti National Park 157–8
Sesamum indicum 108
sex ratio
  adult 193–5
  and effective population size 203
  and genetic drift 203
sexual maturity 179
  age at 105–6
  environmental factors 104, 105
sheep 43
Shettihally Wildlife Sanctuary 12
shifting cultivation 32–3, 144–5
  fires 151
  habitat effects 208–9
  Sholaga tribe 43, 53, 144
shola forest 13, 14
  study area 40, 44, 47, 49
Sholaga tribe 43, 53, 144
shooting 125–6, 168, 169–70
  numbers killed 9
Shorea robusta 15
short grass habitat 62
  browse plants 74
  browse proportion of diet 73
  feeding *70*
  palatability of plants 82
  seasonal use 66, 68
shrub productivity 97
Shwe-U-Daung Sanctuary 18
Sibsagar hills 17
sightings 54
Sigur plateau 13
silver oak 32
  plantations 43, 154, 155
Singhbhum tract 15
slash-and-burn cultivation 144–5
slaughter of elephants 165

sloth bear 52
smuggling 35
soap-nut 149
social structure of elephants 50
socio-economic
    factors in population dynamics 106
    survey 59
sodium
    content of finger millet 134
    content of paddy 134
    requirement 82, 84
soil
    nutrient levels 106
    type and grass productivity 89
Somawathiya National Park 26
sorghum 111
    crop raiding *120*, 123
    potential crop loss 125
Sorghum vulgare 108
sound, high-frequency 215
South Wynad 13
Southeast Asia 18–20
spines 84
sport hunting 7
spotlights 126
spotted deer 51–2
Sri Lanka 24–6, *31*
    browse diet 74
    civil war 37
    feeding habits 70
    religious and cultural associations with man 7
    slaughter and capture 8
    surveys 10
    war elephants 5
stability, ecological 102
stable limit cycle 100–101, 102
stables 4, 5
Stone Age hunters 2
striped hyaena 52
study
    aims 39–40
    methods 54–9
study area
    climate 42–3
    hydrology 40, 42
    land-use pattern 43
    large mammal population 50–2
    location 39, 40
    people 53
    rainfall 42
    satellite image *41*
    temperature 42
    vegetation types 39, 44, *45–6*, 47, *48–9*
sugar cane 108, 114
Sulu, Sultan of 28
Sumatra 26–8, *31*
    Javan transmigration scheme 34
    logging regulations 32
    permanent cultivation 33–4
    surveys 10
Sus scrofa 52

Suvarnavati Reservoir 42, 132
syringe dart 216

Tabin Wildlife Reserve 28
Talamalai
    crop raiding 116
    herd movements 131
    plateau study area 39
    potential crop loss 125
tall grass forest
    browse plants 74
    diet 73
    elephant density 60, 62
    palatability of plants 82
    protein level 67–8
    seasonal movement to 68
Taman Negara National Park 24
Tamanthi Sanctuary 18
Tamarindus indica 71, 149, 150
    plantations 154
Tamil Sangam literature 4
Tamilnadu 13
tannins 84
tea plantations 15
teak 15, 32, 71, 72
    damage by elephants 97
    forests of Burma 20
    plantations 13, 15, 43, 154, 155, *156*
    timber felling 146
Tectona grandis: *see* teak
teeth 81
temperature of study area 42
Tenasserim
    poaching 36
    range 20
Terai forest 16
Terminalia chebula 149
Terminalia tomentosa 146
testosterone 140
Thailand 20–1, *31*
    poaching 36
    smuggling 35
    surveys 10
Themeda cymbaria 48, 69, 83
Themeda spp. 149
    effect of fires 152
    protein level 83, 85
    wet season grazing 73
Themeda triandra 69
thermoregulation 82
thorn forest 60
thorns 84
tiger 52
timber 32
    felling 145–7
    industry and use of elephants 35
    poaching 170
    products of study area 43
Tirupati Hills 4
toxins 84
trade 3, 5

trampling 137
translocation 24, 217–18, 218
   crop protection 215
trees
   causes of damage 100
   cover 153
   crops 108
   cyclical relationship with elephants 100
   damage 57, 107
   effect of fires 152
   productivity 97
   rate of destruction 99–100
trenches 126, *127*, 211
Trikonamadu Nature Reserve 26
Tripura *30*
   political unrest 37
   shifting cultivation 33
trunk 81
   throwing with 137
Tsavo National Park (Kenya) 65, 100, 200
   elephant density 101
   elephant mortality 101
   population crash 106
tusks 2, 4, 35
   absence in southern India 177
   broken 140, *141*
   calcium for growth 82
   goring 140
   hunting in Burma 20
   proportion of bulls with 165
   weight 168, 169
   see also ivory

Ulu Sembakung Reserve 28
Upanishads 2
Uttar Pradesh 16, *30*

Varushanad hills 14–15
vegetation
   primary versus secondary 157–9
   study area 44, *45–6*, 47, *48–9*
   water availability 43
vehicles 126
Vietnam 22, *31*
   American bombing of elephants 8, 37
   elephant meat 36
   war 8, 37
villages *109*
vitamin B12 84
von Bertalanffy functions 54–5

war 8, 37
   uses of elephants 3
Wasgamuwa National Park 26
water
   availability 67–8, 208

   competition for 131–2
   for consumption by elephants 43
   creation of reservoirs 159–64
   effect on vegetation 43
   elephant movements for 131–2
   infiltration rate 106
   limiting factor in home range size 65
   limiting resource 99
   requirements 79–80
   sodium content 84
   table and trees 100
wattle 32
Way Kambas Game Reserve 217
weeds
   bamboo suppression 148
   in cleared areas 144, 145, 155
   invasion after grass cutting 149
   secondary habitats 208
Western Ghats 4, 11, *12*
   hydro-electric and irrigation dams 34
wet season 43
   crops 108
   diet 73
   distribution 62
   grass palatability 82
   habitat use 66
   mortality 200
   quantity of forage consumed *78*, *79*
wild dog 52
wildlife sanctuaries 7
wild pig 52
   crop damage 125, 128
Wilpattu National Park 24, 217
wood
   fuel 43, 146, 147
   hard 145–6
   timber felling 145–7
wood-apple 149, 150
woodland savanna 20
woody
   plants in diet 76–7
   vegetation and elephants' impact 91–7
World War II (1939–45) 7
World Wide Fund for Nature 10, 26
worship 4

Xishuangbanna 18
xylazine 217

Yala Strict Nature Reserve 26

Zea mays 108
   damage area 123
   potential crop loss 125
Ziziphus spp. 71, 149

Schmitt

St. Louis Community College
at Meramec
Library